CAREERS IN
TELEVISION AND RADIO

CAREERS IN **TELEVISION AND RADIO**

Julia Allen

Second Edition

Acknowledgements

The author would like to thank the many people who took the trouble to reply at length and in detail to her requests for information. She also wishes to thank the people who found time to see her and answer innumerable questions, including managers of BBC and independent local radio stations, hospital radio staff, broadcasters, engineers and technicians. Finally, it would not have been possible to write this book without the IBA's yearbook *Television and Radio 1986*, *The BBC Annual Report and Handbook 1986* and the Independent Television Companies Association's *Careers in Independent Television*. Readers should try to obtain copies of these indispensable publications.

First published in 1986 by
Kogan Page Limited
120 Pentonville Road
London N1 9JN
Reprinted in larger format with revisions 1988

The case studies on pages 29, 34, 35, 38 and 46
are reprinted from *Careers in Radio and Television*
by Susan Crimp.

British Library Cataloguing in Publication Data

Allen, Julia
Careers in television and radio.
1. Broadcasting — Vocational
Guidance — Great Britain
I. Title
384.54'023'41 PN1990.45
ISBN 1—85091—538—5

Typeset by DP Photosetting, Aylesbury, Bucks
Printed in Great Britain by
Richard Clay, The Chaucer Press, Bungay

Contents

Introduction

Television and radio in the UK, run as public services and on a commercial basis, provide information, education and entertainment. The British Broadcasting Corporation (BBC), financed largely by the television licence fee, has two television channels, four national radio networks and 31 local radio stations. Independent television and radio companies, regulated by the Independent Broadcasting Authority (IBA), are financed by advertising revenue. There are 15 independent regional television companies, Channel Four and Sainel Pedwar Cymru, its Welsh counterpart, TV-am, Independent Television News Ltd (ITN) and 48 independent local radio stations. Radio Luxembourg and Manx Radio have also been licensed for many years to broadcast in the UK.

The early 1980s was a time of expansion - Channel Four and Sianel Pedwar Cymru went on the air, the BBC and TV-am began breakfast-time programmes and many new local radio stations opened. For the BBC and independent companies that expansion has now peaked; the second half of the decade is likely to be a period of consolidation and staff numbers should be held fairly steady.

Broadcasting, however, will continue to expand, though at the moment it is impossible to say exactly how fast and in which directions. Commercial cable television companies are providing a service to subscribers in a few areas and cable coverage will spread. When direct broadcasting by satellite (DBS) begins it will offer a wide choice of viewing including programmes from neighbouring countries whose satellite 'footprint' falls on the UK. At the beginning of 1987 a contract for Britain's first DBS television service was awarded to British Satellite Broadcasting, which plans to go on the air with three channels in 1990. A great many new independent television production companies are being set up to meet the demand for programmes from Channel Four, cable company operators and the home video market. These companies also produce advertising, promotional, educational and training material. There should be a lot of jobs created and most of them will be in the new sectors; the BBC and the companies regulated by the IBA have lost the programme-making monopoly they once held.

Nobody needs to be told how stimulating and rewarding it can be to work in television and radio, and this book, far from urging its readers to try for a broadcasting career, seeks to present some of the facts as neutrally as possible and may even be thought to have a rather discouraging tone. The really determined and dedicated will, of course, not be put off and they are the ones who will succeed. The book will have served one of its purposes if it makes a few of the thousands who every year send in unsolicited applications for jobs in broadcasting organisations pause and reflect.

Part 1

Chapter 1

The Employers

The BBC

The biggest employer by far among the broadcasting organisations is the BBC which with its domestic and external services is concerned with all aspects of broadcasting. These include news gathering, programme making, selling programmes to foreign buyers, audience research and the development, running and maintenance of the transmitter networks.

The Corporation employs in all some 28,000 people. Vacancies are advertised in selected national newspapers and journals, in the BBC's own publications the *Listener* and *Ariel* - the latter being the house newspaper - and in specialised journals such as the *Stage* and *UK Press Gazette*. Vacancies for certain posts outside London are advertised in the provincial and local press. Much of the BBC's outside recruitment is at assistant level for the Corporation has its own competitive promotion and transfer system, an integral part of which is the attachment scheme. Employees who have applied for promotion or transfer and who have shown particular aptitude and commitment are given the opportunity to work in a new post for a trial period of six months. When this period is up, the employees on attachment compete with other applicants for the post; if they turn out not to be of the calibre the post demands they can return to their old post which will have been kept open.

It can be exceedingly difficult to get a job in the BBC. In a year when 3,318 individual advertisements were issued the Appointments Department received 49,031 external applications, 18,194 internal staff applications and, in addition, there were 50,295 unsolicited enquiries. Information about opportunities in non-engineering categories and about trainee schemes can be obtained from the Head of Appointments. Qualified engineers are recruited regularly for the television, radio, external broadcasting, communications and transmitter groups of the BBC and details of engineer-

All addresses are given in Chapter 6 pages 80–9.

ing opportunities are available from the BBC's Engineering Recruitment Officer. The Corporation has introduced an engineering graduate sponsorship scheme in certain areas, notably in research and design. It has also supported the efforts made by the Engineering Council and the Equal Opportunities Commission to encourage more young women to enter engineering.

The BBC is committed to a policy of equal opportunity in recruitment and advancement for all irrespective of sex, marital status, creed or ethnic origin. This policy also applies to the recruitment of registered disabled persons.

Training

The BBC provides extensive in-house training for personnel who have been in post for some time. Staff attend on-site courses within the different BBC production centres and in the regions and are sent to courses outside the BBC - eg among office staff there is a strong demand for training places in word processing.

The BBC's Engineering Training Department employs 60 lecturers responsible for 90 different residential courses and has fully equipped radio and television training studios, working transmitters and all the major engineering components in the broadcasting chain. The Department is constantly updating existing courses and creating new ones in order to keep pace with developments in modern technology. In 1984, 292 school-leavers (aged 18 plus), 170 direct-entry graduates and 685 BBC staff in post received training - these figures illustrate the scale of the Department's activities.

There are a number of special departmental training schemes for recruits in computing, news, radio sound operations, studio management, television and radio production, engineering and film. These schemes always attract a very large number of applications.

There is a short work-experience placement scheme for school and college students. Priority is given to applicants interested in office work since most of the placement opportunities are in offices. About 30 such placements are made annually.

Each year the BBC runs training courses for overseas broadcasters. Course subjects include journalism, engineering, television production, television production in education, radio production, radio training and radio management.

The Television Training Department runs all the formal production and direction training in the television service. Assistant producers receive practical instruction in film and studio direction.

The BBC External Services

The External Services are intended to provide a link of information and entertainment between the people of Britain and those in other parts of the world, to present news of world events with speed and accuracy and to reflect British opinion and the British way of life. In addition to the broadcast transmissions the BBC sends many recorded programmes to radio stations overseas and provides a worldwide service of English teaching by radio and television.

The External Services employ 2,000 people, very few of whom are based abroad, and 600 of whom are completely bilingual or are people whose first language is not English.

The World Service

The World Service is on the air in English 24 hours a day and the Language Services broadcast to the world in 36 languages. These services broadcast for 727 hours a week and the output includes more than 1,750 news programmes every week.

The Monitoring Service

The Monitoring Service at Caversham Park in Berkshire monitors broadcasts, jointly with US partners, in more than 50 languages from 120 countries. Information and documentation from this joint operation are available in a fast teleprinted news file and daily publication, *Summary of World Broadcasts*. There is a staff of 500 at Caversham, 60 of whom are engineers and 130 of whom are completely bilingual or whose first language is not English.

Both the Monitoring Service and the World Service are part of the BBC's External Services which are funded by direct grant from the Foreign Office.

Ceefax

Ceefax, the BBC's teletext service, is received in 2 million homes in the UK. The service, which is constantly being expanded and improved technically, offers 600 pages of news and information on BBC1 and BBC2 and is available whenever BBC transmitters are on the air. Ceefax is staffed by researchers, subtitlers, engineers and support personnel.

Work Abroad

A number of BBC staff are based abroad - mainly correspondents, engineers and office personnel. All recruiting for overseas posts takes place in London. Home-based staff occasionally get the opportunity to spend short periods abroad when they are involved, for example, in location work or foreign tours.

Local Radio

Local radio is a very important and relatively new part of the BBC's work; the first stations went on the air in 1967. By the end of 1985, 31 stations were in operation in England and the Channel Islands and 85 per cent of the population had access to the service. When the chain is complete 90 per cent of the population will be able to tune in to BBC Local Radio. A station in Chelmsford, Essex, has begun transmission and other areas that will eventually be linked to the chain are Suffolk, Surrey, Warwickshire, Wiltshire, Dorset, Gloucestershire, the Thames Valley and Hereford and Worcester.

The new stations will, naturally, create jobs: for example, BBC Essex will have a staff of approximately 34. Applicants for jobs with BBC Local Radio should be qualified and experienced, though staff receive further training on appointment.

There are three types of BBC Local Radio station: City stations (in London, Manchester and Birmingham) with staffs of 40 plus; A-stations (in Sheffield, Humberside and Leicester) with staffs of 32 plus; and B-stations - the county stations - with staffs of 22 plus. There will be some jobs created because the BBC plans to turn its B-stations into A-stations. Each station is run by a manager supported by a programme organiser, a news editor, a management assistant, an engineer in charge, engineers, news producers, general producers, reporters, programme assistants, secretaries, receptionists and a gramophone librarian.

In Scotland, Wales and Northern Ireland, community stations (not to be confused with Community Radio - see page 21) opt out for part of each day from the Regional Radio Service with local programming (Radios Aberdeen, Highland, nan Eilean, Orkney, Shetland, Solway, Tweed, Foyle, Clwyd and Gwent).

Commercial Activities

The BBC publishes *Radio Times*, the *Listener*, *Ariel*, *BBC Wildlife*, a large number of books, educational material and back-up material for its programmes. It also markets its products: these may be programmes which are sold to foreign broadcasting companies, video cassettes, records and tapes, technical production facilities, microcomputers or articles manufactured under licence. The Corporation also organises exhibitions and commercial courses.

The Open University (OU)

Open University academics and BBC producers work together at the BBC's OU Production Centre at Milton Keynes to produce the audio and video components of OU courses. For the 1985 OU academic year the BBC transmitted about 65 television and 18 radio programmes a week.

The OU Production Centre offers its production expertise and full range of technical facilities to a number of agencies embarking on educational and training projects for schools, colleges and industry.

The Independent Sector

Over 20,000 people work in independent broadcasting and this figure can be roughly broken down as follows: 15,000 are employed by the independent television area contractors, TV-am, Independent Television News (ITN), ORACLE and Channel Four Television Company; some 2,000 work in independent local radio (ILR) and 1,500 in the Independent Broadcasting Authority (IBA). There are also many people associated with independent broadcasting working in independent production companies and ancillary organisations. All jobs are open to both men and women regardless of marital status or ethnic origin.

So great is the number of unsolicited applications for jobs with the ITV companies that vacancies are not always advertised, but when they are it is in the national, local or trade press. If you are thinking of writing on spec to a television company you should be thoroughly familiar with that company's output and state precisely where your abilities and interest lie. Trainee engineers, trainee set and graphic designers and trainee make-up artists will be expected to have some professional qualifications when applying for a post and great care should be exercised in selecting a course. It is a wise precaution to find out what the company you would like to work for thinks of the course you are planning to take.

In most companies there is on-the-job training and courses are provided on contract by certain external institutions. Most technical vacancies are for experienced staff, though some companies have technical training facilities, eg Granada's Engineering Apprentices Scheme. Occasional graduate trainee schemes for researchers and journalists do exist with ITN and other ITV companies and a special project has been set up to give women and ethnic minorities more opportunities in television.

The Independent Broadcasting Authority (IBA)

The IBA fulfils the wishes of parliament in providing television and radio services of information, education and entertainment additional to those of the BBC. It also ensures that they are of a high standard. Its four main functions are: to select and appoint the ITV and ILR companies, supervise the programming, control the advertising and transmit the output.

It has a staff of about 1,500 of whom 250 are based at the London headquarters, 620 in the engineering and administrative centre near Winchester and the remainder mostly in the UK regions. There are seven divisions: the largest, the Engineering Division, is responsible for the operation, maintenance, design and construction of the IBA transmitter system, experimental and development work, network planning and operations, radio wave propagation and planning, engineering information work and technical training.

The other divisions are Television, Radio, Advertising Control, Finance, Information and Administration. The IBA does not employ programme-making staff. It has a general training section which assesses individual needs through a performance review scheme and tries to provide appropriate training in professional, managerial and business schemes. A technical training group provides engineering and related technical training and there is a training scheme for transmitter engineers.

The Independent Television Area Contractors

There are 15 totally separate regional television companies each with its own operating, recruiting and training policies. These are Anglia Television, Border Television, Central Independent Television, Channel Television, Grampian Television, Granada Television, HTV (Wales and West of England), London Weekend Television, Scottish Television, Thames Television, TSW - Television South West, Television South, Tyne Tees Television, Ulster Television and Yorkshire Television (see addresses on pages 86–8). These companies vary considerably in size and employment profiles: the biggest, Thames Television, having a staff of 2,250 and the smallest, Channel Television, employing only 86. All provide programmes for the network and make regional programmes for their area audience.

Central Independent, Granada, London Weekend, Thames and Yorkshire Television - sometimes called 'the big five' - are located in the most densely populated areas of the country and have the highest income from advertising. They make most of the network drama and light entertainment programmes and therefore jobs in any of these companies are likely to be very specialised. The medium-sized companies produce documentaries, quizzes and dramas, as well as local programmes. The smaller companies concentrate mainly on local programmes and offer excellent work experience since they give their staff more involvement in a wider range of tasks than the larger companies.

Channel Four Television Company Ltd

Channel Four, which went on the air in 1982, provides a national service networked to the whole of the UK except Wales. The company employs 280 people and offers a more limited range of jobs than the other independent television companies because it does not make programmes or sell advertising time. There are 40 managers, 56 engineers, 37 commissioning/presentation editors, 7 production assistants, 59 secretarial staff, 5 computer staff, 5 announcers and 71 staff of other categories. Numbers are expected to remain unchanged over the next five years. Most of those who work for Channel Four are salaried employees; freelance or fixed-term contract staff tend to be commissioning editors, continuity announcers or replacements for staff on holiday.

Vacancies are advertised in the national press or in specialist magazines, eg *Marketing Week*, *Computing* or *Broadcast*. Competition for posts is keen; there are usually 100 applications for every specialised vacancy and 500 when a vacancy for an announcer's post is advertised. In addition the company receives about 250 unsolicited applications every month; a large percentage of these are for production jobs which do not exist in Channel Four.

Currently the only in-house training scheme is for technicians, who are given basic training that will enable them to progress to the engineering grades; however, employees are sent to a variety of outside training courses.

Sianel Pedwar Cymru (S4C)

On the Sianel Pedwar Cymru service in Wales the Welsh Fourth Channel Authority schedules some 22 hours a week of Welsh language programmes supplied by outside bodies. In addition, it relays most of Channel Four's 70 hours of programmes either simultaneously or on a rescheduled basis.

TV-am

TV-am, which holds the franchise for breakfast-time television, is a wholly independent company selling its own advertising time and originating its own output. It has country-wide coverage, is on the air seven days a week, and offers a magazine programme consisting of news and current affairs, interviews, sport, items for children, travel information, a weather forecast and features. Weekend programmes show a change of pace and content.

TV-am employs a full range of staff.

Independent Television News (ITN)

ITN is a London-based company owned by all 15 regional television companies, whose main function is to provide the daily programmes of national and international news for the independent television network and a weekday news and news analysis programme for Channel Four.

It has its own studios and employs a wide range of staff including 170 journalists (50 of whom are reporters) and 32 news-gathering electronic camera crews, and also maintains a number of foreign news bureaux.

ITN usually recruits experienced journalists, but occasionally it takes on a small number of graduates with no professional experience for its Graduate Editorial Trainee Scheme. Applicants should hold a good degree and be able to show proof of an aptitude for news writing.

Independent Local Radio (ILR)

There are 48 ILR stations around the country (see pages 84-6) and they vary considerably in size, the largest outside London being Radio Clyde with 80 employees, and the smallest Moray Firth Radio with 20. The London Broadcasting Company (LBC) and Capital employ 161 and 180 people respectively. It is difficult to generalise about ILR as not only is each station an independent company but each has its own strongly local character.

The categories of staff in ILR include managing director, director, station co-ordinator, engineer, technician, producer, head of news, journalist, editor (sports, news, features etc), programme controller, presenter, record librarian, technical operator, and sales, administrative and support personnel. Employment profiles vary from company to company and every station is staffed by a mix of salaried employees and freelances, broadcasters and sales personnel.

ILR has entered a period of consolidation to conserve resources - the rapid growth has peaked and it is unlikely that any new jobs will be created at present. When vacancies occur they are not always advertised because stations have their own contacts on whom they can draw as they receive a very large number of unsolicited applications and there is also a certain amount of inter-station mobility. Occasionally vacancies are advertised in the national, local or trade press (depending on the category of post to be filled) and on the air. Stations tend to show a strong preference for local applicants; they need staff who know the region, its history, culture, social and economic situation, and who can pronounce the local place names.

Training is an individual company responsibility and policy varies from station to station. There is an extremely limited intake of trainees; stations prefer to recruit experienced personnel and give supplementary on-the-job training when it is needed. Some stations use outside consultants to train personnel to handle new equipment and to update the training of sales staff.

Several ILR stations enjoy excellent relations with schools, colleges, Youth Training Scheme organisers and hospital radios and provide work experience for young people.

ILR output consists of programmes especially designed to meet the needs and interests of the local population: news, weather forecasts, traffic and travel information, short features, phone-ins, music and entertainment plus a certain amount of networked material are all broadcast.

The London Broadcasting Company (LBC) and Independent Radio News (IRN)

LBC is different from other ILR stations as it is solely a news and information station. IRN is its wholly owned subsidiary. 161 staff are employed at LBC/IRN; 107 journalists, 17 in administration, 28 engineers and 9 teleprinter operators. Editorial staff, which includes reporters, presenters, producers and researchers, are all journalists. On the editorial side only experienced personnel with at least three years' broadcasting experience are recruited and the company no longer provides any training. Staff numbers are expected to remain fairly constant with enormous competition for the few posts available. If editorial jobs are advertised it is in the *Guardian*; administrative posts are advertised in the relevant trade papers or through agencies. Alternatively applicants may write to the personnel manager or the managing editor for details of staff vacancies.

ORACLE Teletext Ltd

ORACLE provides the on-screen information service, teletext for ITV and Channel Four and the acronym translates as Optional Reception of Announcements by Coded Line Electronics. The output includes news, games and competitions, sports, weather and travel information, a 'what's-on' and television guide, classified advertisements and commercial advertising.

A total of 67 people work for ORACLE Teletext: 10 sales staff, 2 marketing staff, 2 financial staff, 28 researchers, 11 subtitlers, 4 engineers, 4 advertising production staff and 6 secretarial and administrative staff.

Jobs are advertised in the national press, on teletext or through recruitment agencies. ORACLE has been expanding since it became commercialised in 1981 and is expected to continue to do so. The company recruits researchers with a journalistic background and trains its keyboard operators in-house to use the machinery.

Independent Television Companies Association (ITCA)

ITCA employs approximately 100 people who are mainly specialised administrators, secretarial and support staff. It provides a central secretariat to service those needs of the industry that require a centralised approach. There is a programme planning secretariat, an industrial relations secretariat, an engineering department and an advertising copy clearance department.

The ITCA offers a range of training courses for ITV staff throughout the country.

Independent Television Publications

Independent Television Publications publishes *TV Times* and *Look-in* magazines. It has two subsidiary companies - Independent Television Books Ltd and Purselynn Ltd, operating as Independent Television Marketing Enterprises and providing marketing and ancillary publishing services to the parent company.

Radio Marketing Bureau (RMB)
The RMB was set up in 1983 to promote radio as an advertising medium. It also acts as a clearing house for information from UK and foreign radio stations and provides information on radio marketing for all interested parties. Seven people work at the Bureau: one chief executive, three marketing executives and three secretaries. There is no in-house training scheme; the company recruits trained personnel and rarely takes on trainees.

The Association of Independent Radio Contractors (AIRC)
AIRC is jointly funded by the ILR companies. Its functions include representing independent radio to the public and providing a forum for discussion on collective policy between the companies.

Radio Luxembourg

Europe's oldest commercial radio station has been on the air for half a century. It was set up in the 1930s to provide light entertainment. During the war its transmitters were used to broadcast Nazi propaganda, but when the war was over it soon attracted back its huge, loyal audience. In 1948 Radio Luxembourg broadcast its first Sunday night Top Twenty chart.

Radio Luxembourg (London) Ltd is a company wholly owned by Radio Tele Luxembourg (RTL); it is responsible for operating the English Service on 208m and Community Service (local FM programme in Luxembourg) and also represents RTL's interests in the UK. The main departments in the London company are direction and finance, sales, programming and news. There are production studios in London while the DJs and back-up staff are based in Luxembourg. The broadcast teams in Luxembourg keep in touch with the British music scene through the company's head office and studios in London and live link-ups between London and Luxembourg are used for interviews. Vacancies are advertised in appropriate publications, eg a sales vacancy might be advertised in *Campaign*. There is no in-house training but each year Radio Luxembourg welcomes trainee journalists to its newsroom.

The output is not limited to those programmes that can be heard on 208 metres. Weekly shows are made for a number of American stations, there is a two-hour programme of British music broadcast each week in mainland China and a daily English language service broadcast live to Luxembourg and the surrounding border areas of Belgium, Germany and France.

Manx Radio

Manx Radio is unique in British broadcasting. It is a commercial station which started broadcasting in June 1964, long before commercial radio became part of everyday life in Britain. It operates on a licence issued by the British Post Office under UK wireless telegraphy legislation, but the programme content is entirely the prerogative of the Isle of Man authorities. This freedom in terms of programming permits Manx Radio to take sponsored shows. The station broadcasts to all the coastal regions surrounding the Irish Sea. It is on the air from 6.55 am to 1 am every day and its output includes news and current affairs, sport, features, and music that is 'easy listening'. The station is run by a staff of 25: a general manager, five clerical staff, four sales staff, seven news and current affairs staff, four engineers and one production manager, one features producer and two staff presenters. In addition, Manx Radio employs a range of freelance/contract presentation staff in a variety of capacities from daily air work to casual once-a-week programmes. Because of work permit restrictions in the Isle of Man most staff and freelances are recruited locally.

The Services Sound and Vision Corporation (SSVC)

The SSVC provides the armed forces with a variety of entertainment, engineering support and training services both commercially and at the behest of the Ministry of Defence. SSVC Television and the British Forces Broadcasting Service (BFBS), the radio division, are non-profit-making and funded by the Ministry of Defence. All broadcasting staff are civilians.

SSVC Television operates services in Germany and Cyprus and its output consists almost entirely of programmes taken from the BBC and IBA channels under contract. BFBS radio operates in Germany, Cyprus, Gibraltar, Hong Kong, Brunei, Belize and the Falkland Islands; the great majority of the output is original and consists of entertainment, information and educational material. BFBS London does not transmit live but records some 50 hours of speech and music every week; this material is syndicated to overseas stations and received by up to 30 of Her Majesty's ships at sea.

The radio and television services employ approximately 250 people worldwide. Categories of staff employed include, in addition to managers and administrators, producer-presenters, engineers, secretaries and support staff (drivers and messengers etc).

Professional vacancies are advertised in *Broadcast*, the *Daily Telegraph*, the *Guardian* and the trade press. The radio and television services employ experienced, professional broadcasters and trainees although there is no formal trainee scheme; both offer permanent, pensionable employment and short-term contracts.

Cable Television

Cable television has been talked about in the UK for many years but it is only just beginning to operate. At the moment there are a number of 'pockets' where there is a service; these include Aberdeen, Glasgow, Coventry, Swindon, Windsor, the City of Westminster, Croydon, Cardiff, Southampton, the Borough of Camden, Edinburgh and Preston, and cable television is expected to spread gradually throughout the country. The service is run by private companies and financed at the moment by subscription.

If cable television is successful it will create thousands of jobs, for not only will there be an enormous potential market for programmes of all kinds, but the cables themselves will have to be laid and the network operated and maintained.

Further information on cable television can be obtained from the Cable Television Authority and the Cable Television Association, which is the trade association (see page 88). A list of cable operators and cable programme providers can be found in *The Electronic Media Directory and New Media Yearbook* published by WOAC Communications Company.

British Satellite Broadcasting (BSB)

BSB has been awarded the contract for Britain's first DBS television service. The London-based consortium is planning four services on its three channels: a film service, daytime programmes for children, a 24-hour news and events service and an entertainment channel. It hopes to go on the air in 1990 and over the next five years could create up to 25,000 jobs.

Community Radio

Community Radio was to have begun as a two-year experiment in 1986 but at the last moment the government decided to suspend the project. However, in the 1987 green paper on the future of radio the government promised three national commercial radio networks and 'hundreds' of local and community radio stations for the 1990s.

Independent Television Production Companies

There are a great many independent television production companies making anything from full-length television programmes for transmission on the national networks to commercials, publicity, educational and training material. Many of these companies employ a full complement of technical and production personnel. Independent production companies are listed in Kemps *Interna-*

tional Film and Television Yearbook and *The Electronic Media Directory and New Media Yearbook* published by WOAC Communications Company.

Foreign Broadcasting Organisations

A number of foreign broadcasting organisations, such as Antenne 2, The Australian Broadcasting Commission, The Canadian Broadcasting Corporation, Norddeutscher Rundfunk, Radio France, Télévision Française TFI, and the US networks, have offices in London. Key posts in these organisations will be held by foreign nationals but it is possible that there may be a number of support posts open to British subjects. An Embassy should be able to supply the names and addresses of its country's broadcasting organisations running a London office.

Chapter 2

The Jobs

Radio and television offer a huge range of jobs; there are openings for engineers, technicians, journalists, craft workers, secretarial and clerical staff, administrators, accountants, lawyers, librarians, computer staff, sales staff, actors, dancers, musicians, writers and those with creative or artistic talents. All jobs with broadcasting companies are highly sought after; competition is fiercer than in any other sector and employers can be very selective. The only area where there is a shortage of qualified personnel is engineering.

Many people are so keen to get into broadcasting that they will put in for posts they would not dream of accepting in any other sector, in the hope that once inside they will be able to work their way up. There are undoubtedly certain advantages to being inside - you can learn an enormous amount by watching the experts, you can find out what a job involves, you may be able to practise using complicated studio equipment and you are on the spot when an opportunity occurs.

Each post, of course, has its own entry requirements, but as a result of the present job shortage more people are competing for fewer places and graduates are applying for posts for which it is not stipulated that applicants should hold a degree and which offer salaries lower than a graduate normally expects. In a competitive situation good academic qualifications can give applicants the edge, but academic qualifications alone are not enough; employers want evidence of serious commitment and this will show up in such things as outside interests and involvement in voluntary or student activities. But radio and television companies are not full of tea-makers with PhDs because employers are not only interested in academic excellence; they may be looking for people with the manual dexterity to handle complicated equipment, or with creative flair or with good spelling and fast, accurate typing. A personal characteristic they will certainly demand from all applicants for whatever post is the ability to work in a team, as radio and television work is team work. A television play, to take a somewhat extreme example, involves scores of people: the cast, the make-up and wardrobe teams, lighting, sound and camera crews, scene and prop hands, the stage manager, floor manager, production assistant and producer. These are only the people you will find in the studio. Before the production can get to the studio, writers, musicians, legal experts, accountants, scene designers and secretaries will have been involved and when the shooting is over there is work for editors and dubbing mixers.

Television and radio are on the air seven days a week, 365 days a year and some stations broadcast 24 hours a day. Few jobs connected with programme making and transmission are nine-to-five jobs; many involve working unsocial hours and possibly spending periods away from home. No one thinking of working in radio or television should have any illusions about the disruptive effect it can have on family and social life and the toll it can take on health. The presenters and production team of an early-morning programme have to rise at 4am every day, news typists can expect night shift work, evening presenters and studio staff get home at 2 or 3am night after night, and the all-night broadcaster battles home to bed through the morning rush-hour. Deadlines have to be met and enormous pressure can build up, especially in a newsroom; as a result the stress rating for some television and radio jobs is very high.

The very competitive atmosphere of a broadcasting company can be emotionally and physically taxing and in some companies a trainee can expect promotion to assistant after two years, but thereafter all further advancement may be on merit and there are fewer places at the top of the ladder.

Certain traineeships require applicants to have some professional qualifications, others require none; however people interested in working on the administrative side of broadcasting should be professionally qualified, though they need not necessarily have had extensive experience. There are posts for legal experts, for professional accountants, librarians, personnel officers, secretaries, etc.

It is very difficult to give any meaningful information on salaries; they vary from one company to another, change regularly and many salaried staff receive considerably more than their basic pay each month because overtime, regional weightings and shift allowances etc are added on. All that can be said is that salaries in broadcasting compare very favourably with those in other sectors. Many people working in radio or television are members of a union and rates of pay and conditions of work are negotiated by unions and management.

A lot of broadcasters are not permanent salaried staff but are freelances with a contract for a single programme or series of programmes. This is true particularly of the best known personalities in broadcasting, the show hosts, quiz chairmen and actors, many of whom do a lot of work outside radio and television. But other categories of employee also work on contract.

People who apply for a first job in radio or television, particularly if it is a traineeship, are usually in their early twenties. There are openings for school-leavers in craft, clerical or junior posts and on apprenticeship schemes and experienced, mature applicants are often recruited to senior administrative posts.

Certain posts, like that of floor manager or producer, are open only to internal applicants. There is quite a lot of inter-media mobility: staff move from local to national radio, from radio to television, back and forth between the theatre, films and television and from newspaper journalism to radio or television journalism.

One way to present the jobs descriptions that follow would have been to divide them into jobs in radio and jobs in television and then to group radio and television jobs under such headings as artistic, administrative, technical etc. However, that would have been unsatisfactory because many jobs are common to both media and many call for creativity *and* a high degree of technical competence or for administrative *and* artistic expertise. The simple, though not ideal, solution that has been adopted is to present the jobs in alphabetical order. Anyone using this book will realise the obvious fact that camera crews and make-up artists are not employed in radio, and readers interested in, say, library work or sound recording will realise that careers are open to them in both radio and television.

Job titles vary from one broadcasting organisation to another as do precise areas of responsibility and entry requirements. The job descriptions are only indicative; none of them exactly fits a particular job with a specific company.

Accountant

There are two kinds of work for accountants in broadcasting organisations: financial accounting and programme accounting. In the former, staff help prepare and control long- and medium-term company and departmental budgets, provide financial reports and audits, supervise wage, salary and expense payments and manage pension funds. Programme accountants provide management with information on the state of programme budgets and what different departments are spending. This type of accountant needs a thorough knowledge of how programmes are made in order to estimate and monitor production costs, and often works closely with directors. Professional accountants, trainees and juniors are recruited for accounts department work.

Administration

The range of administrative departments in a broadcasting organisation is very wide and includes production planning and scheduling, programme planning, central services, transport and purchasing etc.

Announcer

Announcers (sometimes known as continuity announcers) provide the links between and within programmes. They also, to a great extent, project the company's image, so different companies will be looking for very different kinds of people. All announcers, however, whether they have a regional accent or received pronunciation (standard English), need a well modulated voice, a warm friendly personality and, for television, an attractive appearance. The job, with its aura of glamour, is one of the most sought after in radio and television, but there is more to announcing than introducing or previewing a programme. Many announcers write their own continuity material and if there is a breakdown the announcer has to *ad lib* to fill an awkward gap. The work may include interviewing, reading scripted commentaries, 'filler' pieces, news bulletins and news flashes etc. Most announcers have had an A level or university education and some speech training or theatrical experience. Vacancies are usually for trainee announcers.

Audio Work (see Sound Operator)

Camera Operator

Two kinds of camera are used in television: electronic cameras, which record on magnetic videotape and can be used for live transmission, and film cameras, which are usually used to make commercials and full-length drama features. The use of electronic cameras is increasing, particularly of the lightweight electronic news-gathering (ENG) camera. It is not difficult for those who have acquired film camera skills to transfer them to the electronic camera.

Camera operators are usually recruited at the age of 18 plus and trained on the job. Most applicants will have O and A levels or equivalents and possess some knowledge of electronics, optics, film and television photography. In addition they must have normal colour vision and a good eye for picture composition.

Camera operators' work is tiring; it involves a lot of standing under hot studio lights, the hours are often unsocial and camera crews can be required to work away from base, sometimes in uncomfortable conditions, for long periods.

Electronic-camera Operator

In a studio there may be six electronic cameras, each operated by one person, assisted by camera crew members. Some companies start their trainees operating cameras early on but others make them spend several months operating cranes and moving cables. It usually takes six years to become fully skilled and promotion to the top grades of the profession can take between 10 and 20 years. The electronic-camera operator will usually decide the framing and the composition of the picture, but will be guided by the programme director on how the action is to be shot.

Film Camera Operator
Trainees begin by performing such tasks as pulling focus, loading and charging magazines, checking and cleaning equipment, and recording takes. They learn to operate cameras, and as they acquire skill and experience may specialise in a certain kind of production. Senior film camera operators are responsible for the quality of the pictures both technically and artistically and share with the director major decisions about camera positioning, lighting and how the action will be shot.

Carpenter and Joiner
Carpenters and joiners interpret the set designer's drawings and construct sets and items such as furniture for use on sets. Most of these things are dismantled after use.

Clerk (see Secretary and Clerk)

Computer Personnel
Computers are used for such tasks as stock and cost control, cataloguing and audience analysis. Computer departments recruit data preparation operators, computer operators, systems analysts, programmers and technical and support staff.

Continuity Person (see also Production Assistant)
Television viewers are both observant and critical and delight in pointing out inconsistencies and mistakes. Every detail of hair, dress and story content must be consistent. Sometimes it is one of the production assistant's duties to see that this is so or there may be a continuity person whose responsibility it is to make sure, for example, that if a character sets out in the morning wearing a red coat she returns home in the evening in the same coat. Continuity is also important in radio productions, particularly in serials. In long-running serials personal records are kept of the characters detailing such things as birthdays, wedding anniversaries and how they address other characters in the serial.

Costume Designer
Costume designers work in the areas of television light entertainment and drama. They begin by reading the programme script, then, in liaison with the producer, director, choreographer and set designer, plan the costumes and work out the costume budget.

Costume designers should have a good grounding in the history of costume and etiquette and be creative and innovative while possessing administrative and supervisory skills.

Some television companies have their own stock of costumes, others rely on theatrical costumiers.

Costume design assistant is one of the entry points to this profession and the assistant's duties include arranging fittings, doing research and shopping for fabrics and accessories. Applicants normally hold a degree or equivalent qualification in theatrical or fashion design and have had professional experience.

Craft Posts

Any large company will employ in-house craftspeople to carry out routine installation and maintenance tasks and in a television company there will usually be a number of craft posts. Vacancies are generally advertised in the local press and filled by qualified craftspeople. A few companies recruit apprentices. (See also Carpenters and Joiners, Scenic Artists and Painters, Plumbers, Upholsterers and Electricians.)

Designer (see Television Set Designer)

Disc Jockey/DJ (see Music Presenter)

Director (see Programme Director)

Dresser

Dressers are responsible for the maintenance of costumes and for helping artistes on and off with their costumes. They carry out minor alterations and must therefore be able to sew quickly and neatly. They need tact and sensitivity when dealing with artistes. Both men and women can work as dressers, but people who are under the age of 20 are rarely recruited as dressers.

Dressmaker

Dressmakers work on any kind of costume from a caveman's skin to a martian's spacesuit and are expected to make their own patterns from the designer's working drawings. A dressmaker will have had basic training in pattern making, cutting and dressmaking to BTEC Higher National Diploma level and work experience as a proven dressmaker or with a fashion house or theatrical costumier.

Editor — Film, Videotape, ENG

It is the editor who prepares the final version of a programme. The work demands great attention to detail, precision and creativity and the skills take a long time to accumulate.

Film Editor

Film, unlike videotape, is cut and spliced. The film editor studies each frame of a developed film and decides which shall be removed and where out of sequence shots shall be inserted. Most film editors begin as trainee assistants. Trainees are recruited at the age of 18 or 19; good colour vision is essential and most will have had a broad general education with at least English and maths to O level or equivalent. Promotion to assistant usually comes after nine months. The assistant, who does not make editing decisions, helps the editor by logging film in and out, marking up effects, doing joins, synchronising rushes and keeping in touch with film processing laboratories.

Case Study

Patrick is now working as a *film editor*, a position he has worked towards since his college days. His first contact with television and film production started while he was attending Sheffield Polytechnic School of Art and Design. Although Patrick's main study was fine art, he chose as his subsidiary subjects film and television production.

> On completion of the course, I was accepted by the Royal College of Art, School of Film and Television, to take my master of arts degree.
>
> On graduation, I found myself unemployed. In order to survive I worked during this brief period as a gardener. Months later I was eventually offered a job as a freelance assistant film editor. There were minor problems because of non-union membership, but eventually I got my union membership.
>
> After a short time I moved to another television company as a staff member and remained in the position for nine months. I worked briefly for a commercial company as an assistant film editor, but didn't like it very much so I moved on.
>
> Now I'm a freelance assistant on contract to a television company. I work mainly on drama series, which are really interesting. My work involves editing film, mainly of a dramatic variety, in order to meet the requirements of a pre-arranged programme format laid down by the director. The job itself, though, offers me a great opportunity to be creative and allows me to experiment too. I must bear in mind that I am ultimately responsible to the director, and I must be prepared both to be adaptable and to work under his guidance.
>
> As far as the future is concerned, I would eventually like to be a director myself.

Videotape Editor

As most television programmes are recorded on videotape there is a great demand for videotape editors. The tape is contained on reels and when edit points have been decided the sections of the tape to be used in the production are recorded on to another tape. Editing machines are complicated pieces of equipment but relatively easy to learn to operate; it is not necessary to have a technical background in order to become a videotape editor though many of them start out as television engineers or technicians. The assistant videotape editor does support work for the editor: keeping records and lining up the machines for use and other such tasks.

ENG (Electronic News-Gathering) Editor

ENG editors use the same techniques as videotape editors but work with small cassettes rather than large reels. They have to meet deadlines for news bulletins and are usually working under great pressure. ENG editors are expected to be able to maintain and repair their own equipment as many of them work away from base.

Electrician

Production electricians, also known as lighting electricians, follow the plans of the lighting director and arrange the lamps in a way that will produce the desired lighting effects. They may work in a studio, on location or take part in outside broadcasts. They repair and maintain the apparatus they use.

Engineering

Engineers, technical assistants and technical operators have a vital role to play in broadcasting, but it is beyond the scope of a book of this size to give details of all the jobs available in this vast field. Such information can be obtained from the Training Adviser of the Independent Television Companies Association and from the Engineering Recruitment Office of the BBC.

A few successful graduates in electronics, electrical engineering and applied physics are recruited for direct appointment as engineers.

Operational Engineers

They are responsible for the technical facilities needed for the production and transmission of radio and television programmes. They may be involved in studio work, outside broadcasts, the operation and maintenance of the networks, recording, news-gathering and transmission systems (eg cable and satellite) etc. Some work at base and spend their days sitting at a control desk. Others are out with mobile recording units, at transmitter sites, on location or on overseas assignments.

Specialist Engineers

This category of employee works in research departments developing new techniques and systems and improving existing equipment, in design departments and in capital projects departments.

Technical Assistants
These assistants, when training to become engineers, provide the support force for qualified engineers. They set up, align and maintain broadcasting equipment and in some cases operate it. Applicants for traineeships should be aged between 18 and 26, have normal colour vision and hearing and hold (as minimum qualifications) three GCSEs at grades A, B or C or equivalents in English, maths and physics, a BTEC National Certificate or Diploma in maths, physics or engineering or have studied maths and physics to A level. Many applicants will hold a degree in engineering, electronics or telecommunications.

Technical Operators
A technical operator will be a junior member of a skilled team who helps with the preparation and operation of equipment. Most of the work of technical operators is done in studios, but occasionally they go on outside broadcasts. Applicants should be aged 18-plus, have normal colour vision and hearing and hold at least three O level or equivalent passes in English, maths and physics. Without engineering qualifications technical operators' career prospects can be limited.

Case Study
Robin joined the BBC as a *technical operator* but is now a studio manager.

> I entered as the lowest of the low. I had an Ordinary National Diploma in engineering, the minimum qualification. I had always set my sights on the BBC, I just love working equipment and it never occurred to me that I would do anything else. I had been involved in sound and lighting for amateur dramatics and had projected 35mm films. I'd always been doing technical things with equipment. I think they recognise genuine enthusiasm. They're not taken in by someone who says 'I've done six months in hospital radio' because they think that sounds good. I was taken on as a technical operator and sent on a training course. The work was very specialised and at first I was afraid I'd bitten off more than I could chew, but in fact I soon mastered the job. In the old days there was a combination of disc cutting and taping and the early tape machines were very, very complicated and difficult to use. Now it's just a matter of switching them on. I did a course on the recording side which was a bit of a backwater in a way. I benefited from the attachment scheme and went on four different attachments. I got the chance to become a studio manager through a quirk of history as the jobs of technical operator and studio manager were combined.

Film Assistant (Projectionist)
Film assistants/projectionists operate projection equipment in dubbing and viewing studios and may be involved in dubbing work

– eg combining a number of sound tracks, perhaps those containing music or special effects to produce the final sound track.

Film-Camera Operator (see Camera Operator)

Film Editor (see Editor)

Floor Manager

Floor managers, who have an important and demanding job, are rarely recruited externally; as a rule they start their careers as assistants. They need to be thoroughly familiar with every aspect of television production. One part of their work is to provide the link between the programme director and the studio floor. During studio rehearsals and recordings the programme director normally watches proceedings on television monitors and can speak to the floor manager through a pair of headphones. It is the floor manager who passes on the director's instructions to performers and also gives cues and prompts.

Floor managers co-ordinate and manage everything that happens on the studio floor, making sure that performers know where to stand and what to do, that props are in place and microphones and cameras are correctly positioned, etc. They are in charge of a studio audience when there is one.

Floor managers may work on outside broadcasts and at film locations, doing much the same thing as they do in a studio production. The work calls for stamina, tact, organising ability and great calm.

Trainees are usually aged between 18 and 24, have had a good general education to at least O level or equivalent standard and experience of working with actors. (See also Studio Manager.)

Graphic Design

Graphic designers, working with the production director and set designer, design and supervise the execution of title sequences, credits and other graphic programme material, which can include cartoon sequences, charts of all kinds and such things as documents used as props. The work may be done by hand, with printing equipment or with computers and it calls for great artistic creativity and a range of technical skills. Graphic designers are rarely appointed from external applicants. The graphic design department also employs photographers, photographic assistants and rostrum camera operators.

Graphic Design Assistants

Graphic design assistants work under the designer on all forms of graphic design. They must be able to carry out instructions accurately and neatly and work without close supervision. Trainees are recruited at age 21-plus. They should have good colour vision and hold a degree or equivalent qualification in graphic design.

Graphic Assistants
Their job is to produce a variety of typographic material, for example credit captions or labels, and carry out such tasks as operating caption generators.

Journalism and News Work
News is a very important part of any broadcasting organisation's output. An assortment of people are involved in news collecting, writing and presentation, from the foreign correspondent to the newsroom typist, and all work under pressure; there are strict deadlines, news bulletins are broadcast live and news is continually changing - new stories break and items have to be updated. Some stations are on the air 24 hours a day, there are early morning and late night news bulletins; news work never stops, it involves unsocial hours, weekend duties and night shifts.

Most bulletins contain international, national and local items. Stories come in from various sources including news agencies, stringers (overseas reporters working on a freelance basis), newspaper reporters and from the police. The *news editors* decide on the content of the bulletin and the weighting of the various items that make it up. They decide which items shall be followed up and send a *reporter* to cover the story. The *radio reporter* might go alone with a tape recorder, the *television reporter* would take along a camera crew. The big broadcasting organisations employ *correspondents* who specialise in such subjects as politics or industrial relations, to provide in-depth coverage of important issues. They also employ *foreign correspondents* who are based abroad in major capitals.

News Readers
Also called news presenters, newscasters and anchormen/women are frequently experienced *journalists*. They present the news from the studio, reading some items themselves and linking and introducing stories from other journalists. They may read from a prepared script or they may write their own. They must be thoroughly acquainted with the background to the news and must check such things as the pronunciation of foreign names. They must be able to read at sight without fluffing, as stories may break in the middle of a bulletin, and to vary the pace of their delivery in order to fit in with time signals and call signs.

Journalists
In addition to writing news stories, they also write promotion scripts and press releases. There is intense competition for journalists' posts in broadcasting. Applicants for traineeships are normally

aged between 21 and 25, most hold university degrees or an equivalent qualification and all can show examples of their work: newspaper articles, audio or videotapes.

News Typists

They have the important and demanding task of typing news bulletins and must be able to take dictation directly on to the typewriter and do audio and copy typing. They are trained to operate an electronic distribution system of visual display units. Candidates for news typists' posts should be aged 18-plus, have had a good education and be able to type accurately at 50 words per minute.

Case Study

Hazel has worked as a *reporter* on a local radio station for nearly a year. She first decided on a career in journalism whilst at university. There she had her own chat show on the campus radio station, and also wrote numerous articles for the university paper.

After university, I started the radio journalism course at the London College of Printing. Looking back on it now, I feel that the course was generally good and provided tremendous training as well as enabling me to get on-the-job training in the form of working attachments.

My job as a local radio reporter necessitates an enormous amount of shift work and in view of this, outside work commitments seem invariably to move to second place. Despite this, I enjoy the job tremendously and there is a great amount of variety. It involves a good deal of interviewing as well as writing reports and copy stories. I am always reading news bulletins and often I am sent out to cover council meetings and press conferences. In addition to these assignments, I often do on-the-spot reports, usually live from the scene of the story. These require on-the-spot thought and I admit that at first they were very demanding.

I think being a radio reporter demands great stamina and commitment, but I feel that the job itself has provided me with tremendous scope for a successful career in broadcasting.

I entered radio journalism by taking a radio journalism course rather than gaining training through the newspaper scheme, and I think that this has given me a much wider conception of radio and its requirements, than would have been the case if I had only studied journalism on its own without a radio emphasis. For me the listener is now always paramount, which is of vital importance in radio.

My work is essentially geared towards the presentation of news bulletins. Usually bulletins have a three-minute duration but longer, five- and 30-minute news output does take place, usually at peak listening times during each day.

My job involves finding stories, making phone calls, interviewing, editing interviews (cutting or splicing the tape to acquire the right information and length of time) and reading news bulletins.

It is extremely varied but working towards deadlines does create a pressured atmosphere. Although essentially working shifts, I very often find these overlapping because of the outbreak of important stories which have to be covered. This means working long hours. At the end of each shift, I am responsible for handing over stories to be carried on by the journalist starting the next shift.

For the future, I would like to become a news producer perhaps within the next two years. I am chiefly attracted towards this field because it will enable me to produce longer features and documentaries. Other requirements of such a position are running a news desk with ultimate responsibility for the day's news coverage, a position which would be not only challenging, but extremely demanding and worthwhile.

Case Study

Christine is a *freelance radio journalist.* Here she describes the pros and cons of her job.

Unless you have a thick skin, lots of energy, masses of ideas and endless patience - forget it! Because it could be said that working freelance is like working twice as hard for half the money, and without the staff perks of the organisation. Even in some canteens, the freelance is expected to pay as much as double what staff members pay for the same meal.

If you have something broadcast as a freelance, there's no time to rest on your laurels, you've got to come up with the next idea almost before you've finished the project in hand, otherwise your producer will lose interest. This in itself is a problem (both keeping in touch with the sort of ideas the station wants, and keeping in touch with the producers) because, of course, you are freelance and can't maintain the day-to-day contacts that the staff journalists have with the department. Remember that the continuity between you and the department is always at risk.

Be careful that you get paid a fair rate. If you're an NUJ member, use the NUJ to help you; they have special freelance rates, so if you're a member, make sure you're getting the full NUJ rate, as a lot of producers are uncertain of the rates and if they can acquire something cheap, I'm afraid some of the more unscrupulous ones will. It is also worth knowing that without NUJ membership it is almost impossible to get your work accepted.

Be careful when you go to a producer with an idea that *you* want to expand on, that he doesn't pinch the idea and give it to an in-house journalist to do. Tell the producer the idea, but not the *whole* story, so that you've still got something to pull out of the hat, because often it might be easier for him to get a staff member to do it, so you're only protecting your own interests. Try to take along a certain amount of work that you've already done on the subject, say, an edited interview that can be played to the producer when you meet to discuss the idea. This will make it much easier for him to get an idea of how it will sound as a programme. It may also make him feel more obliged to ask you to do the rest of the work since you've shown initiative by doing this, and haven't just turned up with an airy fairy idea.

On the whole, it's better to ring producers than write to them, because once again it's easier to talk your way into seeing them than waiting sometimes for weeks on end for a reply to a letter. Producers are there, amongst other things, to see people like you so if you can convince them you've got something they can use on their programme, they will want to see you - don't forget it can be quite a headache filling programmes each day or week, so someone who can come up with workable ideas will be very welcome.

Read, read, read, read everything you can lay your hands on for ideas, especially the daily newspapers, but be prepared always to think of a different angle to the story, as there's no point in just broadcasting a newspaper story. Get to know as many of the producers as possible, and always know the format of a producer's programme before you go and talk to him, as nothing irritates a producer more than someone who thinks he can contribute to a programme when he hasn't the faintest idea what goes into the programme.

Try to be as independent as possible, ie if you can afford it, get your own Uher (the standard tape recorder in the radio industry). You can buy them secondhand through HiFi magazines or the various radio stations themselves. It's also very nice to have your own editing facilities, as then you don't take up valuable time on the station's machines. It also means that you can turn round much more work instead of wasting time going to and fro borrowing Uhers, and waiting to use editing rooms.

You will find that being a freelance, what you lose on the swings you gain on the roundabouts. The very fact that you are free to go about rather than be locked into a nine-to-five routine, means that you are in a good position to go around and meet people and also get hold of the more interesting stories. This is simply because you're not in a secure job, but living from hand to mouth, which can make for a much better journalist than one who is cosy and secure, and in many cases deadened by that very security that a staff job gives.

Be organised. Always research your projects as much as possible, make use of libraries, etc, and set aside several days just to collect the material that you need, recorded interviews, etc. Don't start putting it all together until you have the material, all of it, otherwise you will get into a muddle, and never take something up to a producer until it's edited with a brief script. Lastly, be ruthless with your editing; the most they're going to want for a news type item is, say, two to three minutes if that, and for a feature, well this can vary, but probably 15 to 20 minutes at the most. So if you're interviewing someone who's going on a bit, stop him and take him back to the point. Or be brave and once you've got what you wanted from him even if it has only taken two minutes after an hour and a half's drive to get there, turn off the recorder. Remember, you're the one who's got to listen to it all back, and editing is made a lot easier if there isn't too much volume!

Legal Work

In broadcasting there is a great deal of work connected with contracts. Actors, dancers, musicians, show hosts, compères and panelists etc are employed for one programme or a series on a contract which sets out their payment, conditions of service and entitlement to residual rights. Work in a contracts department demands a thorough knowledge of contract, copyright, employment and trade union laws and legislation.

Other fields in which legal expertise is required include video piracy, classification, data protection, libel laws and court reporting.

Librarian

There are a number of posts for graduate chartered librarians, assistant librarians and clerical library assistants. The fields include reference, news, music, archives, film and videotape, photographs, scripts and gramophone records. Librarians perform the normal range of library work: cataloguing, classification, indexing, retrieval etc, and in some instances carry out research. Most of the work is related to programme production. Film and videotape librarians need a knowledge of production, storage and handling techniques and can be asked to carry out simple editing. Music librarians in a small organisation may be required to have a wide general music knowledge. In a very large organisation the work is likely to be specialised.

Lighting Director

Lighting directors decide how to position lights on a set in order to produce the best effect. In a chat show, for example, this would be relatively simple but in a play, lighting is used to create atmosphere and illusion. Lighting directors liaise with set designers, make-up artists and other members of the production team and prepare plans for the *lighting electricians* and *lighting console operators*, whose work they supervise.

Lighting directors are usually recruited internally; they need both creative flair and technical knowledge.

Lighting Electrician (see Electrician)

Make-up Artist

Make-up artists spend much of their time doing corrective work, combing and lacquering hair and powdering noses and foreheads. They also carry out such routine tasks as cleaning wigs and hairpieces and setting up and stocking equipment. The creative side comes from drama and light entertainment productions for which they may have to do elaborate face and body make-up, style hair and wigs or produce special effects such as scars or bruises.

Make-up artists do not always find themselves in the comparative comfort of a studio; location work may take them outside in bad weather. They need a calm, tactful personality for they work with many different sorts of people - actors, politicians, ordinary members of the public - all of whom are likely to be nervous just before going on camera.

Make-up artists usually start as trainees. Applicants for traineeships should be aged 20-plus, have normal colour vision, have had a good general education (A levels in English, art and history would be very desirable) and hold recognised qualifications in make-up and/or hairdressing, and/or beauty culture. Art school qualifications are sometimes acceptable.

Case Study

Amanda decided she wanted to work in television make-up, just after her O levels. From the research she did, she realised that there were usually two ways of entering the field: by taking a hairdressing and beauty training course, or by taking an arts degree. Believing that a vocational course in hairdressing and beauty would be of much wider use, she opted for that and, while studying, made contact with people in the make-up department of a television company. At 20, Amanda was considered rather young to embark on a career as a make-up artist. She decided to apply for another job in the make-up department of a television company and aimed to move upwards within it.

> My job as an assistant in the stock department involves preparing the cosmetics for the artists both for studio make-up and on-location work, as well as ordering the necessary equipment and cosmetics and keeping up-to-date records of the accounts. Although the job is a junior position, I feel that it will eventually help me to achieve the job I want - more so, in fact, than if I had chosen to stay outside the industry until I was a little older. At the moment, I can see exactly what the job of make-up artist involves, what cosmetics and effects are available, and the demands that the job makes on my colleagues. I am now in immediate contact with those who can best advise me and I am really hoping that I will be chosen for the training scheme. This seems to me an advantage I have over those outside the industry.
>
> The career openings in television are few, and in such a competitive field as television make-up it is often a case of getting in wherever you can. Although I can only continue doing a job of this nature for a limited time, it is up to me to do the best I can and with luck I will in the end become a make-up artist.

Marketing Staff

Any broadcasting company that tries to make money from the sale of its programmes employs marketing staff who need a detailed knowledge not only of company output but also of potential foreign markets.

Model Maker (see Visual Effects Designers)

Music Presenter

This is one of the most sought-after jobs in radio, and stations receive hundreds of unsolicited applications and demo tapes. There is, however, much more to it than 'disc spinning': for one thing the position carries considerable responsibility as the DJ is more often than not the station's image figure. A music presenter has to be a very good all-rounder; in addition to technical experience (music presenters in local radio handle their own studio equipment) the job calls for a wide general background including a grasp of current affairs, an easy microphone manner, a pleasant voice and the ability to think quickly and to *ad lib* when necessary. Music presenting includes a lot of routine work - running orders have to be listed, programmes timed and records logged for royalty payments.

Case Study

Paul is a *music presenter*. After obtaining a science degree at Manchester University he decided to go into industry and he gained broadcasting experience working for a hospital radio station in Newcastle. This was an extremely professional set-up, broadcasting to 10 hospitals in the area. Paul's regular job enabled him to travel extensively in the UK and wherever he went he got to know the local radio station.

> My hospital radio work was excellent training. I was determined to become a music presenter on a local station and sent out hundreds of letters and audition tapes, which were met with rejection upon rejection. Eventually one station offered me a series of training sessions and, as a result of these, I landed myself a freelance job on an overnight programme.
>
> After this I was soon offered another programme. Meanwhile I was keeping an eye open for any openings within the company. I was offered holiday relief work which I managed to fit into my annual leave from my other job in industry.
>
> A few months later I was offered a full-time presenter's job on a contract basis and I left my job in industry. My contract was renewed and I was able to gain extra training in the features department.
>
> Looking back, I have no regrets about taking a degree, or indeed about working for a short time in industry. In fact I think the experience gained outside radio has been to my advantage. My advice to anyone trying to get into radio is 'keep at it'. There are bound to be moments of despair as the rejections pour in, but determination is the only way to succeed. Competition for music presenters' jobs is extremely tough - there are thousands of would-be disc jockeys around the country, but there's only room for the good ones.
>
> There's no such thing as a typical day in radio. My schedule depends on how many programmes I'm responsible for in any one week. Generally I have a three-hour afternoon programme six days a week. I usually arrive

at 9.30 and spend the first hour opening mail and making arrangements for it to be answered. Then I concentrate on the content of the afternoon's programme, choosing the records and researching. Once the running order has been planned (the listing of records and the timing of the programme) I log the records for royalty purposes. After lunch, I usually spend the time before going on the air scanning the music press for interesting stories. Sometimes in this period before my programme begins I have to read advertisements or promote the station's activities.

My programme runs from 3pm till 6pm. When it's over I clear away the records and usually set off home by 6.30pm.

News Work (see Journalism)

Performers

Performers (actors, dancers, musicians, stuntmen/women etc) work on a contract basis. They are hired through agencies or may apply independently to audition. The BBC has a number of staff appointments for *musicians*, in the orchestras or in the *BBC Singers*, for which auditions are held.

Acting

Acting is a notoriously overcrowded profession. Only 25 per cent of Equity's 30,000 members are working at any one time and 4 per cent can expect to be out of work for more than 12 months at a stretch. Most radio and television work is in one-off productions, though of course there are long-running serials. Jean Alexander, named 'Television Performer of the Year' by the Royal Television Society, and known to millions as the curler-crowned Hilda Ogden of *Coronation Street*, wrote:

> I did not go to drama school and I gained my first professional experience in travelling theatre. I found myself an agent and he got me my first television part. The contract was for a week. Fame did not arrive overnight and I had to look for other work. When the part of Hilda Ogden came up I had to audition for it.

Personnel Work

Personnel work in a broadcasting organisation is similar to that in any other company or institution. It includes such tasks as recruiting, training, job grading, salary administration, record keeping and industrial relations.

Photographer (see Graphic Design)

Plumber

Plumbers are employed on sets which need working plumbing.

Presenter

Presenters 'front' (or host) current affairs programmes, quizzes, games and chat shows etc and they are usually well known public personalities in such fields as sport, journalism or the theatre. Presenters' jobs are not advertised; presenters are normally approached and offered a contract for a programme or a series.

Case Study

Robert Robinson has been a broadcaster for over 30 years. He started out as a journalist and worked on the *Sunday Chronicle*, the *Sunday Graphic*, the *Sunday Telegraph* and the *Sunday Times*. On radio he has been a presenter of the early-morning news magazine *Today* and chairs *Brain of Britain* and *Stop the Week*. On television he has produced a number of documentaries and chaired *Points of View*, *Call my Bluff*, *Ask the Family* and *The Book Programme*.

None of this will come as a surprise to anyone who has given the matter any real thought.

I wanted to be a journalist, but I don't know why I wanted to be one. So the impulse was an accident. I got the job in spite of, or perhaps because of, wearing a bow-tie and carrying a walking-stick. So that was an accident too.

Then there was a newspaper strike in 1954 and a nice girl had the bright idea of cobbling together some TV programmes in which the journalists deprived of their papers could say what they couldn't write. I knew her from Oxford, so she invited me on, and that was an accident.

I wasn't any good, but neither were the others. Then because I'd been on once I was on twice (you'll find the axiom in Euclid), and because twice, three times.

The next feeling I had was that appearing on television was what more or less everyone did - this allowed me to appear without really noticing that that was what I was doing. This was the biggest accident of all. By the time I started noticing I was on television it was too late for me to so completely balls it up that anyone would decide I ought not to be on it.

Of course, I'm only talking about people who appear on the medium, not those who direct and produce. You need to be told how to do the latter, but I can't see how you can 'train' for the former - it's a quirk of temperament which fits you for it, and a quirk of temperament which makes you want to do it, and if you are fully invested of these two conditions I don't honestly think you even need to practise.

Anyone who needs this information won't benefit from it, and anyone who doesn't could have written it for himself.

Producer

Many people confuse the roles of producer and director. In some productions the same person takes on both roles, but when the jobs are separated it is the producer who heads the production team and who will probably have originated, or contributed to, the idea on which the programme is based. In addition the producer manages the programme budget and the scheduling of rehearsal and recording/shooting and has a say in the selection of actors/participants and members of the production team. Every programme has a producer and most producers specialise. For example, the radio talk on current affairs needs a producer with a good grasp of the subject and sound political judgement; a television wildlife programme might need a producer with the organisational skill and experience to get a large production team and expensive equipment to the Sahara.

Most vacancies are filled by internal applicants. There are a few traineeships for graduates wishing to become producers, assistant producers, researchers or script writers in radio or television.

Production Assistant

Production assistants work on a particular programme from start to finish, providing support services for the programme director. They attend all programme planning meetings, take notes of the decisions made and see that the required action is taken. If during rehearsal the production director decides to change the script, the production assistant notes the changes and retypes the script. Many production assistants do the work of the continuity person (see separate entry). In the final run-through of a television production the production assistant sits with the programme director in the control room and instructs the camera operators over the talkback system, and in both radio and television productions the production assistant times recordings or takes with a stopwatch. The production assistant will provide any information needed by those engaged on post-production work, eg editing.

Production assistants also work on live programmes, eg news, concerts, sporting events, and the atmosphere in the control room during transmission can be very tense. The job requires excellent organisational skills, attention to detail, calm, initiative and the ability to manage and get on with all types of people.

Most production assistants are recruited as trainees and many vacancies for traineeships are filled by people already in broadcasting. Applicants are usually in their early twenties, many are graduates with secretarial experience and all have had a good general education.

Programme Director

The programme director in television is in charge of the shooting of a programme and the direction of performers and technical crews. In a play, there is detailed shot-by-shot work; the director places the actors and props, decides on lighting effects and camera angles etc, consulting the technical experts, and discusses the interpretation of roles. When the shooting has been done the director supervises post-production work such as editing and sound dubbing. In a live broadcast, eg a news bulletin, the programme director follows a running order, selects pictures from those offered by the camera operator or from videotape and relays instructions to the presenters. In radio drama the director rehearses the actors, selects sound effects and after recording supervises the post-production work of editing and dubbing.

Most vacancies are filled by internal applicants with substantial production experience.

Projectionist (see Film Assistant)

Property Staff

Property staff (also known as 'props') look after all the movable items on a set, anything from an armchair to an ashtray, and the work can be strenuous. Trainee property staff should be aged between 18 and 45 and be physically fit. They need no special educational qualifications, though CSEs or equivalents in English, maths and technical drawing would be useful, as would a clean driving licence.

Public Relations Officer (PRO)

Public relations officers present the company to the public. They prepare publicity material, liaise with the community, promote programmes and issue press statements. Training in journalism is good preparation for work in a PR department.

Publishing Work

Many broadcasting companies produce a house organ, an annual report, literature to accompany certain programmes and a number of books. There are also the large-circulation periodicals such as *Radio Times* and *TV Times*. Publications departments employ journalists, editors, secretaries, clerks, proof-readers and book designers.

Reporter (see Journalism)

Researcher

Behind nearly every successful programme there will have been a hardworking, competent researcher, who has sought out the practical elements of, say, a story and put them into a form which can be incorporated in the programme. News, news-magazine and current affairs programmes require extensive collaboration from researchers who are expected to contribute ideas for the programme and prepare the material for particular items. Researchers line up and interview contacts, find suitable people for interviews and write a script or treatment for the programme presenter. They may have to trace pictures in newspaper archives, sequences in old newsreels or recordings in sound archives. General research tasks could include looking up bibliographical details for a book programme, testing consumer reactions to a new product for a food programme or finding participants for a quiz or panel game. Specialist or academic research for particular programmes or series is normally contracted to outside experts.

There is stiff competition for research jobs, and applicants are usually in their twenties, hold a good degree and have had some kind of media-related experience, eg journalism.

Rostrum Camera Operator (see Graphic Design)

Sales Staff (independent/commercial sector only)
Those radio and television companies whose revenue comes from the sale of advertising time employ sales staff – *sales co-ordinators* negotiate the sale of airtime, take bookings and see that 'slots' are filled; *sales (or marketing) executives* are responsible for attracting new business; *sales research staff* carry out and interpret market research; *traffic staff* monitor the make-up of commercial breaks and arrange for the receipt and delivery of advertisements. Trainees for these posts are recruited externally and applicants should have good A levels, a degree or equivalent. Brand selling or advertising agency experience would also be useful.

Scenic Artists and Painters
Scenic artists and painters work under the set designer and paint sets, backcloths, gauzes. They often have to fake effects, eg a marble floor. Some of the work of a scenic artist is very skilled, and applicants for such posts need a degree in fine art painting as well as a thorough knowledge of styles in architecture, painting and furniture. Painters do less specialised work.

Script Editor
Script editors work in drama departments. One of their tasks may be to act as intermediary between the writer and producer and some script editors commission new work. The original idea for a serial usually comes from one person but a number of writers will be needed to produce episodes in a uniform style, sometimes over many years, following a storyline which the script editor may have prepared. Script editors generally have a literary background and have worked in the theatre, reading and reporting on unsolicited scripts.

Secretary and Clerk
There is an exceedingly wide range of jobs for secretaries and clerks in broadcasting: they are employed in every department. In the

different programme departments secretaries could specialise in drama, documentaries, music, light entertainment, features, talks, current affairs, religious or educational broadcasts; there are also secretarial posts in personnel work, management services, finance, legal work (including contract and copyright work), publications, sales, public relations and engineering. The list is long and the work is interesting and can provide a rewarding career. However, many people see secretarial posts as a spring-board to more demanding senior jobs, particularly in production, and for this reason a lot of graduates join broadcasting organisations as secretaries. It is true that many vacancies for production posts are filled by internal applicants, but such posts require specialised knowledge and competition for them is very tough indeed. All staff are expected to remain in their first post for a reasonable length of time before applying for a transfer.

Secretaries may work for one person or for several, alone or in groups. They need accurate shorthand (80/100 wpm) and typing (30/40 wpm) and good educational qualifications, common sense, initiative, adaptability and a good telephone manner.

Clerks do such jobs as handling mail, photocopying, printing and filing. Those who can type (30/35 wpm) or have an aptitude for figures have more opportunities open to them.

Case Study

James is now a music studio manager and a balance engineer for live music broadcasts. He had no formal qualifications but decided that if he could get in to broadcasting on the ground floor he might be able to work his way up; he began as a *clerk*.

> I joined the BBC as a clerk and worked in a library of tapes. At my interview they said 'We're not offering you a career, we're offering you a job'. It was up to me to read the notice boards and see what was going. I worked for two years in the tape library and during that time got to learn what my present job involved. I heard about a traineeship and applied for it. I was in competition with all the other applicants, up before the boards. It may have been a matter of luck the way I landed my present job; the *Third Programme*, as it was called then, was just beginning to broadcast music all day and they needed more people. I'd always been interested in music - I'm a singer and a player, amateur of course - but nowadays, with more people competing for fewer jobs, things have got a lot tougher and I suppose only those with music degrees or a diploma from one of the colleges would get shortlisted.

Set Designer (see Television Set Designer)

Setting Assistant (see Stagehand)

Sound Operator

Sound operators are employed in both radio and television and their work requires technical skills and creativity.

In a radio studio duties include: sound balancing and mixing, tape recording, disc playing and selecting sound effects. In a television studio *sound technicians* see that studio equipment is working properly and that microphone boom arms are positioned so as to remain out of camera shot. They work in concert with the camera operators.

There is a sound control room where technicians monitor the sound signals and feed music or other sound effects into the programme and balance the sound. All sound technicians do a lot of pre- and post-production work. They edit taped speech and music and select, or in some cases devise, special sound effects.

Sound technicians work on outside broadcasts, eg matches and state occasions. In the outside broadcast (OB) unit, which is a large vehicle equipped like a sound studio, sound signals are mixed and monitored. Staff are also responsible for rigging up equipment and testing the quality of lines.

Applicants for traineeships should be aged 18-plus, have normal hearing and colour vision and have had a good general education.

Case Study

Alan always intended to follow a career in music and is now a *sound recordist*. He became a trainee after successfully completing a five-year engineering apprenticeship. As a trainee he performed mainly minor tasks in recording and transmission and was often asked to find pieces of music for drama productions.

> My previous musical interests were obviously of use in my work. Most important of all for me was the fact that while in this position I came into close contact with the rest of the production team and this provided invaluable experience.
>
> After two years as a trainee, I decided to leave the company and accept promotion offered elsewhere. My work now became essentially OB (outside broadcast) which meant that I was working on location away from home for long stretches. Looking back on the move now, I think it provided me with marvellous experience.
>
> With my additional experience, I returned to my old company in the capacity of sound recordist. I can't really imagine doing another job. My work provides a great amount of variety and I look on it not only as a promotional progression, but also as an artistic one.
>
> There are some highly pressurised moments but everything about the job is challenging. It's varied and I find myself meeting a wide range of people every day.
>
> My advice to any would-be sound recordist is to be as flexible as possible, resilient, a little flamboyant and, most important, dedicated.

Stagehand

Stagehands, also known as *setting assistants*, usually work in small teams. They erect the scenery in a studio or on location and dismantle it when shooting or transmission are finished. Most television scenery consists of large plywood flats which must be assembled according to the designer's plans. The work is strenuous.

Trainee stagehands should be aged between 18 and 45 and be physically fit. They need no special educational qualifications, though CSEs or equivalents in English, maths and technical drawing would be useful, as would a clean driving licence.

Stage Manager

Stage managers organise outside rehearsals, which usually take place in large public halls, order and move props and note any changes made to the script during rehearsals. They also mark out the rehearsal room floor and prompt. Stage managers have all had experience of theatre or television.

Studio Manager

Studio managers have an important and demanding job that calls for creative flair, technical competence, organising ability and a cool head. They have to see that the radio producer's/director's instructions are carried out and this involves setting up the studio for recording or transmission, checking equipment, adjusting sound balance controls, monitoring sound quality, running tapes and discs at the right moment and co-ordinating the activities of the people participating in the programme, who might be actors, musicians, a studio audience, people called for interview etc. Studio managers usually start as trainees and then specialise, eg in music broadcasts, current affairs or drama. They may work in a studio or an outside broadcast unit, on live or recorded programmes. The atmosphere during live transmission can be very tense. (See also Floor Manager.)

Supply Operative

Supply operatives collect and transport costumes and accessories. They are attached to the wardrobe department.

Technical Assistant/Operator (see Engineering)

Television Recording Operator

Television recording operators control film and telecine equipment and videotape machines. They must understand the capabilities of the equipment and appreciate the artistic requirements of the production on which they are involved.

Applicants for traineeships should be aged 18-plus, have normal colour vision and hearing and have had a good general education.

Television Set Designer

Set designers not only work on sets for plays; every television programme from a panel game to a studio discussion has a set which has been designed to create a certain ambience. Designers work with producers and directors, they must understand the technical processes of production and have a feeling for how the content of the programme should be interpreted visually. The work requires knowledge of the history of art and architecture. Many television sets are viewed in close-up and from different angles so a great deal of planning goes into them; designers have to take into account the positioning and movement of performers, props, cameras, camera cables, microphone booms and lighting. They usually construct a scale model of a set for use in programme meetings and produce simplified architectural drawings from which costings can be made. When a set design has been agreed plans are sent to the workshop and construction begins.

Television set designers almost invariably begin as assistant designers. Applicants for traineeships should be aged 21-plus and hold a degree or its recognised equivalent in interior design, art and design, stage design or architecture.

Transmission Controller

Transmission controllers send the broadcasting organisation's programme output to the transmitters. This output might originate in a number of different places and it is sent via land line or other links to local transmitters. Transmission controllers sit at a presentation mixer console on which they can monitor the picture that is going from point of origin to transmitter, the picture going from transmitter to home television screen and the opening sequence of the next programme, and they press the controls to bring in each programme on cue. The work involves a great deal of planning and combines periods of intense activity with periods during which nothing much happens. Transmission controllers need to be able to think and act quickly in the event of technical failure as they must not leave viewers with a blank screen. Most vacancies are filled by internal applicants. Applicants for assistant transmission controller traineeships are usually aged between 18 and 21, they need normal colour vision and good hearing and should have had a good general education; many hold A levels or a degree.

Upholsterer

Upholsterers work under the set designer and make upholstery and soft furnishings. There are few vacancies.

Vision Mixer

A television programme may be made up of pictures that come from a number of different sources, eg from a camera in the studio, from pre-recorded videotape, from a telecine machine or from a slide. Vision mixers receive a signal from the director telling them when to cut from one picture source to another producing a smooth sequence of images. The vision mixer sits at a console into which all the picture sources are fed; this console can produce special effects, eg dissolving or wiping, to make the transition from one scene to another either interesting or unobtrusive. Vision mixers work in the studio or on outside broadcasts. The job calls for quick reactions and an excellent sense of timing. Trainees are usually recruited from among those already working in television; they need normal colour vision and good hearing. Most applicants are in their early twenties and have had a good general education.

Visual Effects Designer

Visual effects designers construct and carry out all sorts of visual effects, such as a 'Fire of London' or a rocket launch. They use scale models, made by model makers from a variety of materials such as wood, plastic, papier maché or fibreglass. Visual effects staff need a good working knowledge of sculpture, model-making, painting, optics, pyrotechnics, together with an understanding of the principles of physics, chemistry and electricity. Applicants for the post of visual effects design assistant should be graduates and experience in films or the theatre will stand them in good stead.

Wardrobe Work

Wardrobe stock-keepers issue, maintain and index costumes for television productions. *Wardrobe operatives* perform such tasks as assembling and labelling garments and packing them up for despatch.

Writer

Most of those who write for radio and television are freelances. There is seldom full-time work, except for journalists, but many authors find that writing for radio and television can be an interesting and well paid sideline. It is specialised work with its own techniques and it is advisable to learn these either by reading books, or taking a course, on the subject. Occasionally it is possible to break into script writing when an unsolicited script is accepted, but most writers employ a literary agent who, of course, knows the market. Writers are not always asked to produce original work; sometimes they are asked to contribute episodes for long-running serials or make adaptations of novels or short stories.

Case Study

Sue Limb collected her share of rejection slips when she was starting out as a *writer*; however, she found that one job almost certainly leads to another.

> Tortured teenage poems and satirical sketches formed the bulk of my early writing at school. At 13 I wrote a prose epic in which I bore Richie Benaud's illegitimate son and went off to live in the Australian outback. After such a feverish fantasy life, reading English at Cambridge seemed somewhat humdrum. But *Footlights* provided me with contacts which were to be useful later.
>
> My first published work was *Captain Oates: Soldier and Explorer* (with Patrick Cordingly for Batsford), for which I collected several rejection slips, as I did for *Up the Garden Path*, a comic novel now available in paperback. Indeed, just before it was accepted my agent rang me and said, 'I think we'd better give up with this one and put it in a drawer.'
>
> Rejection was also a feature of my early contact with the BBC. I sent them two plays on spec and they were rejected. A few years later I had an idea for a comedy series, wrote the pilot, and sent it in. They were interested enough to ask me to rewrite it - *nine times*. At this point I withdrew hurt. About a year later I had a better idea, for a series based on the home life of the Wordsworths, *The Wordsmiths of Gorsemere: an Everyday Story of Towering Genius*, and this needed only one rewrite.
>
> Once launched with the BBC, the rest is easy. They will ring up and ask me to do compilations, for example, and the Light Entertainment Department is now very sympathetic to my ideas. I have done masses of work for Schools Radio, and even won a Sony Award for *Big and Little*. I've also been one of Robert Robinson's guests on *Stop the Week*.
>
> I would like to write for TV as it would be interesting to try and manage without words.

Chapter 3

Getting Started

Intense competition makes it very difficult to get started on a career in television or radio and here we look at some of the approaches you can make and that other people have made.

Hospital and campus radio (see below) and disco DJ work all provide the opportunity to learn how to use equipment and practise microphone technique. Community radio (see page 26) will offer similar opportunities. Find out what is going on in your area and how you might become involved.

Pirate radio stations come and go; it has occasionally been said that experience gained on a pirate radio station can provide an entrée to broadcasting, but both the BBC and IBA take a strong line against illegal stations which pay no copyright fees and steal news bulletins that legal organisations have spent a great deal of time and money collecting, and may be reluctant to employ applicants from this area.

Remember that local radio is indeed *local* and whenever possible employs people with local ties and local knowledge. Many people get a foot in the door by doing voluntary, temporary or part-time work for a local station; a marginal job, like manning a switchboard on Sundays or replacing a receptionist who is on holiday, can lead to other things.

If you are lucky enough to be taken on and given the chance to learn some of the ropes you will be expected to show commitment by sticking at it for some time. So ask yourself, when the offer of one evening's work a week for a hospital radio station comes up, if you will be prepared to go on week in, week out, after the novelty has worn off. Remember that working in radio and television involves unsocial hours and your personal life has to take second place.

Applying

Whether you go after an advertised post or simply send in an unsolicited letter of application you are going to have to write a letter and draw up a curriculum vitae (CV). It is very important that these make a good impression because broadcasting organisations receive hundreds - in the case of the BBC, thousands - of letters.

Letter of Application

At the risk of stating the obvious, here are some guidelines:

- ☐ Make a rough draft of your letter of application to be sure that it contains all essential points.
- ☐ It does not matter if you cannot type your letter, but you must write neatly and legibly.
- ☐ Use good quality, preferably white, writing paper and a matching envelope.
- ☐ Address the letter to the Head of Appointments or Chief Personnel Officer or Head of the Recruiting Department (or some other such person).
- ☐ If you know this person's name (it may have been on a brochure or leaflet that you have seen) use it. If you write 'Dear Mr/Mrs Johnson' sign off 'Yours sincerely'; if you do not know the name write 'Dear Sir or Madam' and sign off 'Yours faithfully'.
- ☐ Print your own name under your signature.
- ☐ If you have any doubts about spelling or grammar show your letter to a teacher or tutor or a reliable friend or relative (it is probably a good idea to do this in any case).
- ☐ Keep a copy of your letter for reference.
- ☐ If you are writing in answer to an advertisement mention where you saw it.

Find out all you can about the organisation you are applying to (eg do not write on spec to Channel Four for a production job!) and show that you are familiar with its output. Make your own area of interest clear, for example if it is music say what kind, if it is technical state your qualifications and experience. It is not enough simply to say you would like to work in radio or television. Keep your letter short and make your CV as comprehensive as possible.

Curriculum Vitae

Type your CV (or have it typed) if you possibly can. It is worth paying to have a CV typed professionally; a good typist will know how to set it out and might be able to advise on content and phrasing. A CV should include:

- ☐ Full name and address
- ☐ Date of birth
- ☐ Schools attended
- ☐ Examinations passed (dates and grades)
- ☐ Any other honours won at school or college

- ☐ Any particular position of authority held at school, eg school captain
- ☐ Training courses/colleges attended and qualifications gained (dates and grades)
- ☐ Previous jobs held or any other experience (names of companies and dates)
- ☐ Names and addresses of two referees. One of these should be a previous employer or someone who has personal knowledge of your abilities
- ☐ Personal interests and hobbies
- ☐ Languages - if you can speak or write a foreign language mention it here
- ☐ Driving licence. Mention if you hold a clean, current licence.

Interview
When you are called for interview you may feel very nervous and find it difficult to collect your thoughts when asked even quite simple and obvious questions. It is a good idea to think out what you would say in answer to these questions:

- ☐ Why have you decided to try for a career in radio/television?
- ☐ What made you apply for this particular job?
- ☐ What makes you think you will be good at this job?
- ☐ What attracts you about this job?
- ☐ How would you like your career to develop/What would you like to be doing in five years' time?
- ☐ Why do you want to leave your present job? (If you already have a job)

Accepting a Job
Before you accept a job make sure you know exactly what position and conditions of employment you are being offered.

Contract of Employment
A contract of employment exists as soon as someone offers you a job (even verbally) at a certain rate of pay and you accept. Your employer is required by law to give you written details of your contract within 13 weeks of your starting work. These details are:

- ☐ job title
- ☐ pay
- ☐ how you are paid (weekly, monthly etc)
- ☐ hours of work
- ☐ holiday entitlement and pay
- ☐ length of notice
- ☐ disciplinary and grievance procedures
- ☐ pension rights
- ☐ any requirement to join a specific trade union.

If you are not given a copy of your contract within 13 weeks of joining a firm, you should ask for it. The contract of employment is a legal document, so make sure that you keep it in a safe place.

Hospital Radio

Hospital radio stations - nothing whatsoever to do with the National Health Service - are run by dedicated amateurs in hospitals up and down the country. One or two of the bigger ones are registered charities, but more generally they are funded by a grant from a hospital amenities committee or from 'friends' of the hospital, by money raised through an appeal or by the efforts of a local club or organisation that donates the proceeds of a charity concert or coffee morning. In some hospitals there may be just one person running a very limited service, perhaps an hour of record requests once a week; in others, there is a large team and varied output with local news, local sport, talking books, talking newspapers, interviews, religious services and music.

Hospital radio attracts a number of people who see it as a way of getting into professional broadcasting. It provides a lot of useful experience; volunteers learn how to handle equipment, put together a well-balanced programme and build up a relationship with an unseen audience. They also discover, when the novelty wears off, that the work calls for commitment - as does professional broadcasting - and may involve unsocial hours and sacrifices in social and family life. A good many people drop out after a while.

A hospital radio service may enjoy very good relations with the local radio station, its presenters sometimes taking part in local radio broadcasts or spending time at the station in order to gain experience.

Case Study

Mike runs a hospital radio station, to which he devotes 40 hours of his own time each week, and although he is a regular contributor to a local radio programme, he has never had any thoughts of becoming a professional broadcaster; he enjoys his full-time job as an engineer.

> I've always been interested in radio and electronic devices. I built a crystal set, then a valve amplifier and gradually worked my way through as enthusiasts do. I also became interested in tape recording and joined the local amateur tape recording club where I got quite a lot of experience recording different kinds of music: pop groups, folk singers, orchestras. It gave me great satisfaction. One of my jobs was with a company making television equipment and broadcasting equipment and so when, about 15 years ago, I was asked to help run our hospital radio station I had acquired quite a lot of useful knowledge.

I was surprised to see how crude the system was in those early days - obviously I wasn't expecting anything on the lines of the BBC - but there was no tape equipment at all; all we had were two turntables, a microphone and a pair of headphones, which meant that we couldn't use things like a signature tune or taped inserts.

Gradually, over the years, we have been able to improve our equipment. I find it a very interesting challenge to design and build what we need. Most of the things are not available commercially, they're specialist parts. Of course I do all the maintenance work too. I enjoy designing the equipment and as much as anything sitting back and watching it work. As I operate the station I can decide how to spend what funds we receive. I know whether we need A rather than B.

I've always been interested in music but it had never been my intention to broadcast. I was pushed into the presentation side at the deep end. Our station presenter died very suddenly one Saturday morning and we had a programme to do the following Tuesday. There was no one else in the team; it was me or nothing. Well, it was sink or swim. I was nervous at first and it took me quite a long time to settle down, but I'm perfectly comfortable doing it now and I enjoy the presenting side of the work very much. I'm helped by a team of people who collect the requests for me on Sunday afternoon; this releases me to do the technical jobs.

I love the pleasure patients show when I meet them. I've been here quite a long time; a lot of people know me, or if they don't they see my lapel badge and they come up to me and say: 'Dad was thrilled to bits when he heard his request last week. He'd like you to play another this week.'

Sometimes I go up on the wards and collect requests for a local radio programme which complements our record request programme very well. Our programme, of course, can only be heard in the hospital and it consists of requests made by people in the outside world for the patients; the local radio programme is made up of requests from hospital patients for their families, visitors and medical staff and it can be heard all over the county. It goes out weekly and all the main hospitals in the area take part in turn.

Campus Radio

A number of universities, polytechnics and colleges of further and higher education run courses on broadcasting-related subjects (see pages 62-8) and some of them have studios in which students can learn to use equipment. Certain other universities have a campus radio, run by students, originating its own output and broadcasting to the university audience within its limited service area. It is not possible to give an assessment of the various campus radios because they are very much the product of those who run them and the student population changes every year.

On a good campus radio you can learn to handle broadcasting equipment, develop microphone techniques and discover how hard it is to fill airtime with good, original material. Local radios are often willing to help campus radios by providing extra training and allowing students airtime.

Case Study

Joanna has become a professional broadcaster; she found campus radio a good training ground.

> I took a degree - English and education - it was a three-year course and I graduated in 1982. I was at a university that has its own radio station. It's run entirely by students on a fairly professional basis and is very well kitted out. It broadcasts to the campus which is all on one site. Listening figures are quite high. As it's on the air from 8am to 2am non-stop there's plenty of opportunity to get involved - there's a lot of airtime to fill! In order to be allowed on the air you have to supply the committee with a demo tape and if they think it's all right you can have some time. You can do whatever sort of programme you like, but most of the output is music-based. There's a lunch-time magazine programme that provides opportunities for people interested in news. It's not always very professional sounding; on the other hand it's an excellent way of getting to know the ropes. Learning to operate the mixing desk, the gram decks and the cassettes will stand you in good stead if you go into local radio because a lot of things will be familiar to you.
>
> A campus radio exists as much for the people doing the broadcasting as for the listeners. It provides wonderful opportunities. It's very much a training ground where you can have a go at being a DJ. We're talking about an in-house programme which provides music - and most students want to listen to music every day - a little bit of magazine programme here and there and some sort of light entertainment as well.
>
> I did two weekly programmes. I did the chart programme - you know, like the *Top 40*, running down the new charts and playing most of the discs. With a friend, I also did a programme called *Grot Slot*; we played dreadful records and old sketches. Basically it was just light entertainment. But the real value of doing something like that is the technical practice you get. Campus radio is like local radio: you operate the desk, you do it all yourself, you play in all your own records. It's exactly the same as local radio.
>
> I was also involved in student television, which was a bit ropey but quite good fun. I took certain options in my degree course. I made an educational television programme.

Local Radio

Many people feel that local radio is more approachable than national. BBC and Independent Local Radio station managers are always on the look-out for new talent of all kinds. Some ILRs organise local music contests such as Essex Radio's 'Band Search '85' or the 'Best Band in Wales' competition organised by the Welsh ILR stations and Metro Radio's 'Track to the Top'. Other stations simply audition demo tapes. Station managers receive a great many unsolicited demo tapes and a good rule to observe when making a demo tape is *keep it short*. An all-rounder's tape might contain a two-minute news reading (this would be a piece of about 250 words), a well-structured interview lasting about three minutes and a sample of how you introduce music. *In all 10 minutes maximum.*

Local radio stations are permanently in need of material and are glad to be put on the track of local news, sports news, news from schools and colleges, people to interview and places to visit.

There are good opportunities for freelances, those who have another job, often full-time, but who have a regular weekly slot to present, eg a gardening, cookery, specialist music programme, or a programme for a particular age or ethnic group. Make your ideas known to your local station; you may not be given your own airtime but you could be invited to contribute to an existing programme.

Case Study

Jo is a *programme assistant* with a BBC local radio station: she planned her entry to her chosen career very carefully.

> I went to the university with the intention of getting into broadcasting, which is the only thing I've ever wanted to do. I got involved in campus radio not only because it would look good on my CV but because I actually wanted to; I enjoy doing that sort of thing. Towards the end of my final year I decided to do a secretarial course because otherwise I thought I'd have to put in for one of those trainee schemes that literally thousands of people apply for and eight people get a place. I didn't honestly think I would be one of the eight. The course I chose was very tough, it was a diploma for personal assistants involving law, economics and personnel management. I don't think I would recommend people to do such a tough course if they just want to be a secretary and move on from there. All you really need is good typing and shorthand.
>
> I wanted to work in local radio and being a secretary is a good way of starting off. You get to know the environment, and what goes on and how things in local radio work. And you don't do boring secretarial work. I'd imagined I'd be typing letters for a year, but I was doing nothing of the sort. It may be a bit different if you're secretary to the general manager. General production secretary, or news secretary, those are the jobs to go for if you want to move on, because you're researching interviews, you're setting up guests, you're organising outside broadcasts, you're also doing some technical stuff - basic editing, you're helping to get the news bulletins put together, putting the reports on to cartridge. Also you're in a situation where you've got great opportunities for practising, for getting to know the desk, for practising in the evenings, bringing in records, talking into the mike. You're in a good position to sit in on programmes, watch them going out and learn how it's done. If you were a secretary in London you might not get anywhere near the equipment and you might be trapped.
>
> You've got to take these opportunities; there's no point in sitting around for a year then going to your boss and saying: 'Well, I've been a secretary for a year and now I want to be a programme assistant' if in

that year you've done nothing. I learnt how to use a portable tape recorder so I was doing a few interviews, I was going to the market each week to find out the price of cauliflowers. All these things, if they are of a reasonable standard, can go out on the air which is marvellous. But at the end of the year, even if you've done a lot of these things and impressed people, when a job comes up you've got to compete for it with eight or nine other people. I'd been a secretary for just over a year - they like you to stay for at least a year, to show some commitment - when it just happened that a programme assistant's job cropped up. I put in for that job, along with a lot of other people, and was offered it as a six-month attachment[1] and at the end of six months I had to be interviewed again and compete with eight other candidates. I eventually got the job, but it was not easy - you can get stuck. Once you're in as a secretary you are a staff member, you're in a position to hear about internal jobs and to hear about things on the grapevine and to qualify for attachments.

I'm a programme assistant now and I don't know where I want to go from here. I'd planned everything up to this stage. Obviously I'll be staying around for quite a bit longer. You've got to stay around to gain experience and you have to show commitment; it's no good trying to rush things because you'll just come a cropper. I don't want to flit from one thing to another trying to be a star.

You don't have to be a programme assistant for very long - it depends on how good you are - perhaps three or four years would be the maximum, then, if you were taking the conventional route, you'd go for a producer's job in local radio. A programme assistant is very much a junior producer; you do the same work but have less responsibility. If the programme assistant acts as producer you get paid what is called 'extra responsibility rate'. When I do my Sunday programme I'm paid as a producer because I actually produce, set up and present that programme and the same thing happens when I produce an outside broadcast. I'm not terribly sure that I want to become a producer because I feel there would not be all that much more of a challenge and I think that in a year's time I'll be ready for a new challenge.

The Lucky Break

Some people get into broadcasting almost by chance and launch themselves in a single lucky break. Ted Moult wrote:

I'm a farmer and if you'd told me 30 years ago that I was going to have a parallel career as a broadcaster I'd have thought you were mad. Back in 1957 I saw an advert in the *Radio Times* for contestants to take part in *Brain of Britain*, which in those days was called *What Do You Know?* I had to audition and got a contract, initially, for a single programme. Then the producer of the radio programme put on a television version called *Ask Me Another* and invited me to take part. Each week three *Brain of Britain* winners played an *ad hoc* team of say, three lawyers, politicians or journalists. I was asked to be in a team of three farmers.

[1] The BBC's attachment scheme is explained on page 10.

> Well, that led to other panel games, quizzes and documentaries, I suppose I was what's called a 'natural' because I was offered acting parts. I became a member of Equity and I've appeared in a couple of television plays and three pantomimes. Over the years I've worked with the BBC and most of the Independent Television companies. One day I read somewhere that they were looking for a new Dan Archer so I phoned *The Archers* office and asked if I could audition. I failed that audition but the producer offered me the part of Bill Insley and I enjoy that very much. Dan's well into his eighties now and Bill's a much younger man. I'm still a farmer but I get a lot of fun out of my entertainment and publicity work and my charitable activities.

The BBC and Independent Broadcasting companies are always in need of contestants for radio and television games and competitions. Look out for advertisements or write to the producer of a show you would like to take part in and ask if you can audition. It is a good idea to see what the contestants go through and as most contest programmes have a studio audience you can apply for tickets.

Part 2

Chapter 4

Courses, Awards and Training Schemes

A number of institutions offer courses on broadcasting-related subjects; however, the broadcasting organisations, while acknowledging that many of the courses are of a good standard, do not officially recognise them. The organisations which recruit trainees put them through a house training scheme. If you are thinking of taking a course and hope to work for a particular company on completing it, you could check with the company's recruiting manager that the course has a good reputation.

A degree is always a good recommendation, but it need not be in media studies. Employers look for experience - for those seeking a first job this may be experience gained in school or college productions or publications - evidence of commitment and interest in a precise area of work.

The list of courses below is a basic guideline; for further information consult CRAC *Graduate Studies* (Hobsons Press), CRAC *Directory of Further Education* (Hobsons Press), *British Qualifications* (Kogan Page) and the *Handbook of Degree and Advanced Courses* (National Association of Teachers in Further and Higher Education). When you find a course that interests you full details can be obtained from the institution offering it.

Postgraduate Courses

City University, Graduate Centre for Journalism Studies, 223–227 St John Street, London WC1V 0HB
MA/diploma in radio journalism

Cornwall College of Further and Higher Education, Trebenson Road, Poole, Near Redruth, Cornwall TR15 3RD (0209 712911)
Postgraduate/post-experience diploma in radio journalism

Croydon College of Design and Technology, Barclay Road, Croydon, Surrey (01-688 9271)
Postgraduate diploma in film and television animation

Highbury College of Technology, Department of Journalism Studies, Cosham, Portsmouth PO6 2SA (0705 383131)
Postgraduate diploma in radio journalism (full-time)

Lancashire Polytechnic, Preston, Lancashire PR1 2TQ (0772 22141)
Postgraduate diploma in radio and television journalism (1 year)

London College of Printing (a constituent of the London Institute), Elephant and Castle, London SE1 6SB (01-735 8484)
Postgraduate diploma in radio journalism (1 year full-time)
Postgraduate diploma in design and media technology (3 terms full-time)

London Institute of Education, 20 Bedford Way, London WC1H 0AL (01-636 1500)
MA in film and television studies for education

Middlesex Polytechnic, The Admissions Office, 114 Chase Side, London N14 5PN
MA/postgraduate diploma in film and television studies (1 year full-time)

Polytechnic of Central London, 309 Regent Street, London W1R 8AL (01-486 5811)
PhD/MPhil in film studies
PhD/MPhil in mass media

School of Communication, 18–22 Ridinghouse Street, London W1P 7PD
Postgraduate award in radio journalism

Royal College of Art, Kensington Gore, London SW7 2EU (01-584 5020)
PhD/MA in film and television (3 years full-time)

Saint Martin's School of Art, 107–109 Charing Cross Road, London WC2H 0DU (01-437 0611)
Postgraduate certificate in film and video (1 year full-time)

Sheffield City Polytechnic, Pond Street, Sheffield S1 1WB (0742 20911)
MSc/diploma in film studies (2–2½ years part-time)
MA in communication studies (1 year part-time)

South Glamorgan Institute of Higher Education, Western Avenue, Llandaff, Cardiff CF5 2YB (0222 551111)
Postgraduate diploma in media studies (1 year part-time)

Sunderland Polytechnic, Langham Tower, Ryhope Road, Sunderland SR2 7EE (0783 76231)
MA/diploma in communication studies (2–3 years part-time)

University of Bristol, The Senate House, Bristol BS8 1TH (0272 303030)
Postgraduate certificate in radio, film and television (1 year full-time)

University College at Cardiff, Centre for Journalism Studies, 34 Cathedral Road, Cardiff CF1 9YG (0222 44211)
Postgraduate diploma in radio journalism studies (1 year full-time)

University of Kent at Canterbury, Canterbury CT2 7NZ (0227 66822)

PhD/MA in film studies

University of Leeds, Leeds LS2 9JT (0532 431571)
PhD/MPhil in television

University of London, Goldsmiths' College, Lewisham Way, London SE14 6NW (01-692 7171)
Postgraduate diploma in communications technology (1 year full-time/ 2 years part-time)

University of Stirling, Stirling FK9 4LA (0786 3171)
PhD/MLitt in film and media studies

First Degree Courses

Christ Church College, North Holmes Road, Canterbury, Kent CT1 1 QU (0227 65548)
Classified combined BAHons in radio, film and television studies with education or music (3 years full-time)

College of Ripon and York Saint John, York Campus, Lord Mayor's Walk, York YO3 7EX (0904 56771)
BA, BSc DipHE modular courses including drama, film and television

Coventry (Lanchester) Polytechnic, Priory Street, Coventry CV1 5FB (0203 24166)
BAHons in communication studies (3 years full-time)

Dorset Institute of Higher Education, Wallisdown Road, Wallisdown, Poole, Dorset BH12 5BB (0202 524111)
BA/BAHons in English and media studies
BAHons in communication and media production (3 years full-time)

Glasgow College of Technology, Cowcaddens Road, Glasgow G4 0BA (041-332 7090)
BA in communication studies (3 years full-time)

Harrow College of Higher Education, Watford Road, Northwick Park, Harrow, Middlesex HA1 3TP (01-864 5422)
BAHons in applied photography, film and television (3 years full-time)
BA in photographic media studies (2 years part-time)

King Alfred's College, Sparkford Road, Winchester SO22 4NR (0962 62281)
BAHons in drama, theatre and television studies (3 years full-time)

London College of Printing (a constituent of the London Institute), Elephant and Castle, London SE1 6SB (01-735 8484)
BAHons in film and video (3 years full-time)
BAHons in media and production design (4 years sandwich)
BAHons in visual communications - photography, film and television (full-time)

Manchester Polytechnic, All Saints Building, All Saints, Manchester M15 6BH (061-228 6171)
BAHons in design for communication media (3 years full-time)

Middlesex Polytechnic, The Admissions Office, 114 Chase Side, London N14 5PN (01-886 6599)
BAHons in applied photography, film and television (3 years full-time)

New University of Ulster, Coleraine, County Londonderry (0504 4141)
BAHons in media studies (3 years full-time)

North Cheshire College, Fearnhead, Warrington, Cheshire WA2 0DB (0925 814343)
BA ordinary degree combined studies: media and communication (3 years)

Polytechnic of Central London, 309 Regent Street, London W1R 8AL (01-486 5811)
BAHons in film, video and photographic arts (3 years full-time)
Polytechnic of Central London, Faculty of Communication
BAHons in media studies

Polytechnic of Wales, Pontypridd, Mid Glamorgan CF37 1DL (0443 405133)
BAHons in communication studies (3 years full-time)

Queen Margaret College, 36 Clerwood Terrace, Edinburgh EH12 8TS (031-339 8111)
BA in communication studies (3 years full-time)

Ravensbourne College of Art and Design, Walden Road, Chislehurst, Kent BR7 5SN (01-468 7071)
BAHons in visual communications design (3 years full-time)

Sunderland Polytechnic, Langham Tower, Ryhope Road, Sunderland SR2 7EE (0783 76321)
BA/BAHons in communication studies (3 years full-time)

Trent Polytechnic, Burton Street, Nottingham NG1 4BU (0602 418248)
BAHons in communication studies (3 years full-time)

Trinity and All Saints College, Brownberrie Lane, Horsforth, Leeds LS18 5HD (0532 584341)
BA/BScHons Public media courses taken in combination as part of degree (3 years full-time)

West Surrey College of Art and Design, Falkner Road, The Hart, Farnham, Surrey GL19 7DS (0252 722441)
BAHons in photography, film and video animation (3 years full-time)

BTEC Awards

Brunel Technical College, Ashley Down, Bristol BS7 9BU (0272 41241)
BTEC National Certificate in audio-visual studies

Epsom School of Art and Design, Askley Road, Epsom, Surrey KT18 5BE (037 27 28811)
BTEC National Diploma in audio-visual studies

Dewsbury and Batley Technical and Art College, Halifax Road, West Yorkshire WF13 2AS (0924 465916)
BTEC National Diploma in audio-visual studies

Harrogate College of Arts and Adult Studies, 2 Victoria Avenue, Harrogate, North Yorkshire HG1 1EL (0423 62446)
BTEC National Diploma in audio-visual studies

London College of Fashion, 20 John Prince's Street, London W1M 9HE (01-629 9401)
BTEC (Board for Design) Higher National Diploma in theatre studies and specialist make-up (full-time)

Longlands College of Further Education (Middlesbrough), Douglas Street, Middlesbrough, Cleveland TS4 2JW (0642 248351)
BTEC National Diploma and National Certificate in audio-visual studies

Mid-Cheshire College of Further Education, Hartford Campus, Northwich, Cheshire CW8 1CJ (0606 75281)
BTEC National Certificate in audio-visual studies

North-East Wales Institute of Higher Education, Connah's Quay, Clwyd CH5 4BR (0244 817531)
BTEC National Diploma in audio-visual studies

Ravensbourne College of Art and Design, Walden Road, Chislehurst, Kent BR7 5SN (01-468 7071)
BTEC Higher National Diploma in communications engineering (2 years full-time)
BTEC Higher National Diploma in television programme operations (2 years)

South Thames College, Wandsworth High Street, London SW18 2PP (01-870 2241)
BTEC (Board for Design) Higher National Certificate in audio-visual studies
BTEC National Diploma in audio-visual studies
BTEC National Certificate in audio-visual studies

City and Guilds of London Institute Awards

The C&G 6996 Foundation Course in multi-media communications technology is offered at:

Longlands College of Further Education (Middlesbrough), Douglas Street, Middlesbrough, Cleveland TS4 2JW (0642 248351)

Norwich City College of Further and Higher Education, Ipswich Road, Norwich NR2 2JL (0603 660011)

Watford College, Hempstead Road, Watford, Herts WD1 3EZ (0923 41211)

College Awards

Croydon College, Fairfield, Croydon, Surrey CR9 1DX (01-688 9271)
Advanced certificate in film and television

Darlington College of Technology, Cleveland Avenue, Darlington, County Durham DL3 7BB (0325 67651)
International diploma in journalism (full-time)

Derbyshire College of Higher Education, Kedleston Road, Derby DE3 1GB (0332 47181)
Diploma in film studies (2 years part-time)

Guildhall School of Music and Drama, Barbican, London EC2Y 8DT (01-628 2571)
Certificate in prop making (full-time)
Certificate in scene painting (full-time)
Certificate in stage management (full-time)

Gwent College of Higher Education, Allt-yr-yn Avenue, Newport, Gwent NPT 5XA (0633 51525)
Diploma in film (documentary/animation options)

Havering Technical College (Hornchurch), Ardleigh Green Road, Hornchurch, Essex RM11 2LL (040 24 55011)
Certificate in film studies

Highbury College of Technology, Department of Journalism Studies, Dovercourt Road, Cosham, Portsmouth PO6 2SA (0705 383131)
Diploma in radio journalism (1 year)

LAMDA London Academy of Music and Dramatic Art, Tower House, 226 Cromwell Road, London SW5 0SR (01-373 9883)
Technical certificate in drama (stage management)

Lancashire Polytechnic, Preston PR1 2TQ (0772 22141)
Diploma in radio and television journalism (1 year full-time)

Middlesex Polytechnic, 114 Chase Side, London N14 5PN (01-886 6599)
Certificate in stage management and technical theatre

RADA Royal Academy of Dramatic Art, 62–64 Gower Street, London WC1E 6ED (01-636 7076)
Diploma in stage electrics
Diploma in stage management

Shirecliffe College, Shirecliffe Road, Sheffield S5 8XZ (0742 78301)
Diploma in media studies

Solihull College of Technology, Blossomfield Road, Solihull, West Midlands B91 1SB (021-705 6376)
Certificate in audio-visual design

South-East Derbyshire College, Field Road, Ilkeston, Derbyshire DE7 5RS (0602 324212)
Certificate in media studies (full-time)

Sunderland Polytechnic, Langham Tower, Ryhope Road, Sunderland SR2 7EE (0783 76231)
Diploma in film and television studies (2 years part-time)

College-Based Courses

Amersham College of Further Education, Art and Design, Stanley Hill, Amersham, Buckinghamshire HP7 9HN (024 03 21121)
Pre-entry journalism course

Chippenham Technical College, Cocklebury Road, Chippenham, Wiltshire SN15 3QD (0249 50501)
Pre-vocational media studies

College of Technology, St Brycedale Avenue, Kirkcaldy, Fife KY1 1EX (0592 268591)
Foundation course in media studies

Darlington College of Technology, Cleveland Avenue, Darlington, County Durham DL3 7BB (0325 67651)
Course in radio journalism

East Surrey College, Gatton Point, Redhill, Surrey RH1 2JX (0737 72611)
Pre-entry journalism course

North Herts College, Cambridge Road, Hitchin, Hertfordshire SG4 0JD (0462 2351)
Course in media and communications study

West Bromwich College of Commerce and Technology, Woden Road South, Wednesbury, West Midlands WS10 0PE (021-556 9954)
Course in television and audio production

Training for Radio Journalists

The Association of Independent Radio Contractors (AIRC) and the National Union of Journalists (NUJ) set up The Joint Advisory Committee for Radio Journalism Training (JACTRJ) which, in December 1985, issued revised guidelines for colleges on the basic minimum requirements to be met by postgraduate diploma courses in radio journalism. JACTRJ does not seek to impose standards on colleges but to ensure that students receive a sound grounding in the essential skills of radio journalism and are fully equipped, on completion of their training, to take their place in a radio newsroom as well-rounded professionals.

Practical Training

According to the JACTRJ guidelines this should include the following:

1. The spoken as opposed to the printed word. How to communicate information clearly, succinctly and *accurately* by the

spoken word. Writing for radio, including bulletin stories, cues and voice reports. Correct English usage and pronunciation.
2. A thorough grounding in, and a critical understanding of news values.
3. A thorough knowledge of the full range of news sources and an awareness of the importance of research.
4. Interviewing techniques and an awareness of the different types of interview situations.
5. Bulletin editing.
6. The use of actuality and the preparation of short news packages, features and documentaries.
7. The use of portable tape recorders, microphones, mixers and other studio equipment. Practical tape editing.
8. Voice training to a level acceptable for broadcasting purposes.
9. News reading.
10. A form of shorthand designed to achieve a minimum of 80 wpm and keyboard operation to a level of 30 wpm.

Professional studies

These should aim to provide students with a diversity of background knowledge essential to them as professional journalists and to develop in them a critical understanding of the role and responsibilities of journalism in society.

They should include the following:

1. The structure and functioning of *public administration*, including central and local government, parliament and local councils, the political parties and the political process.
2. *Trade unions and employers' organisations* and other interest groups. Industrial relations.
3. *The development of an informed awareness of major contemporary social issues*, eg housing, education, unemployment, inner city deprivation, race relations etc.
4. *Journalism and the law*, including the structure of the judiciary and court reporting – eg the treatment of rape cases, privilege, contempt, libel and defamation, electoral law, the Official Secrets Act and other relevant subject matter.
5. *Media studies*, including the structure and ownership of the British media and the place and role of the media in society.
6. *Journalistic ethics*, including such issues as privacy versus the public interest, surreptitious recording, fairness and impartiality, the issues involved in editing interviews and selecting quotes, the checking of sources, techniques of news gathering, investigative journalism etc.

It is not intended to suggest that these various areas of study should necessarily be dealt with separately and in an academic mode. Rather they should wherever possible be integrated with practical work so as to highlight their relevance and to achieve an effective and meaningful interpenetration of the theoretical and the practical.

The tutor input based on a direct personal experience of broadcasting or, where relevant, journalism should amount to no less than 40 per cent of the total tutor time, ie two days in five. Colleges must have studio facilities, editing equipment and portable recorders sufficient in number to cater for their student intake.

Assessment

JACTRJ does not wish to suggest that all college courses should be identical or should adopt standardised methods of assessment. Each institution should feel free to develop its own style within the guidelines set out above. However, it is crucial to the raising of standards that colleges should be rigorous in their assessment of students' knowledge and skills. Continuous monitoring of each student's progress by his tutor throughout the course must be a key element in the process of arriving at a final judgement on whether the student reaches the required standard in every field. The assessment should include personal motivation and initiative as well as progress in the development of professional skills.

However, in certain specific areas there should be formal tests which the student should be required to pass. These should take two forms:

1. A written examination to cover the following –
   (a) bulletin and cue writing and a voice piece;
   (b) journalism and the law, public administration and other topics touched on under 'professional studies' above. Particular stress should be placed on the achievement of a high standard of examination work in law for journalists. A merely average performance should not be regarded as sufficient. The equivalent of a 60 per cent pass-mark should be the minimum requirement.
2. A practical work project including the following –
   (a) a five-minute news bulletin prepared and read by the candidate;
   (b) a three-minute news interview;
   (c) a two-minute *vox pop* or human-interest interview;
   (d) a current affairs documentary or feature programme of at least 15 minutes' duration.

Students should be required to pass in all major aspects of the course.

The assessment of the practical programme work should involve at least one external examiner, ie someone who contributed less than 5 per cent of the course teaching and who has had at least five years' continuous full-time experience in radio.

The written examination papers should be approved by the external examiner, who may also require to see the answer papers.

The final certificate should indicate a special merit level which should be the equivalent of an overall passmark of at least 70 per cent.

The following institutions' courses, which place a strong emphasis on practical skills, have been assessed by JACTRJ and received 'recognition': Cornwall College of Further Education and Higher Education, Highbury College of Technology, Lancashire Polytechnic, London College of Printing, Polytechnic of Central London Faculty of Communication.

JOBFIT (Joint Board for Film Industry Training)

JOBFIT was set up in May 1985 by the Association of Cinematograph, Television and Allied Technicians (ACTT) - and two major employers' associations, the British Film and Television Producers' Association (BFTPA) and the Independent Programme Producers' Association (IPPA). The Advertising Film and Video Producers' Association (AFVPA) has now also joined the scheme as a sponsor.

The idea was to create the first systematic, industry-wide training scheme covering all ACTT film grades, for new entrants into the freelance film-making sector, eg art department assistant, assistant script-supervisor, assistant boom operator, clapper/loader, second assistant editor, assistant sound recordist, third assistant director etc. The training will be like that of an apprenticeship - trainees will be attached to various film productions over a two-year period and given supporting technical training. It is hoped that up to 50 people will be trained per year, starting in March 1986. Trainees will be taken on in groups of 12 at intervals. Successful candidates will be contracted to JOBFIT for two years, subject to regular reviews and assessment. On satisfactory completion of the two-year programme, trainees will receive ACTT membership in a junior grade of their chosen specialist area. JOBFIT trainees will be supernumerary to ACTT crews in all instances.

The cost of running JOBFIT for the first two years (until May 1987) has been met by the Greater London Training Board. Training costs and trainees' weekly allowances will be met from an industry-wide levy.

Applying for a JOBFIT Traineeship

Write for an application form when you see the scheme advertised. Note that in autumn 1985, 2,300 people applied for the 50 places available. If you have any queries about JOBFIT please contact the Secretariat.

Chapter 5

Recommended Reading

Books on radio and television pour off the presses - technical manuals, memoirs, biographies and autobiographies, histories, surveys, 'inside stories', research papers, 'how-to' guides ... the list is endless. If you are thinking about a career in broadcasting you should try to read as much as you can, not only on your own subject but also on such general questions as broadcasting in the 1990s.

It is also important to watch and listen to as wide a variety of television and radio programmes as you can - sample the whole range of output - and it is also a useful exercise to compare your own assessment of programmes with that of the critics.

Books and Pamphlets

Action Stations! BBC Local Radio, BBC (1979)

Aitken, Jonathan, *Officially Secret*, Weidenfeld & Nicolson (1971)

Aldred, J, *Manual of Sound Recording*, Argus Press (1978)

Alkin, Glyn, *Sound Recording and Reproduction*, Butterworth, Focal Press (1981)

Alkin, Glyn, *Television Sound Operations*, Butterworth, Focal Press (1975)

Alvardo, M and Bascombe, E, *Hazell, the Making of a Programme*, British Film Institute

Amos, S W, *Principles of Transistor Circuits*, Butterworth (1981)

Amos, S W, *Radio, Television and Audio Technical Reference Book*, Butterworth (1977)

Annan Report. An ITV View, Independent Television Companies Association, ITV Books (1977)

Anwar, M, *Ethnic Minority Broadcasting*, Commission for Racial Equality (1983)

Anwar, M, *Who Tunes in to What? A Report on Ethnic Minority Broadcasting*, Commission for Racial Equality (1978)

Aries, S J, *Dictionary of Telecommunications*, Butterworth (1981)

Aspinall, Richard, *Radio Programme Production: a Manual for Training*, UNESCO (1971)

Ayers, R, *Graphics for Television*, Prentice-Hall (1985)

Baehr, H and Ryan M, *Shut Up and Listen! Women and Local Radio*, Comedia (1984)
Baddeley, W Hugh, *The Technique of Documentary Film Production*, Butterworth, Focal Press (1979)
Baggaley, Jon, *The Psychology of the Television Image*, Gower Press (1980)
Baker, Richard, *Here is the News*, Frewin (1966)
Barnhart, Lyle D, *Radio and Television Announcing*, Prentice-Hall (1950)
BBC Annual Report and Handbook
BBC Ceefax, BBC (1984)
BBC Monitoring Service 1939–79, BBC (1979)
BBC an Outline of its History, Organization and Policy, BBC (1982)
BBC Pronunciation: Policy and Practice, BBC (1974)
BBC Radio for the Nineties, BBC (1983)
BBC Television Centre London, BBC (1982)
Belson, W A, *Television Violence and the Adolescent Boy*, Saxon House (1978)
Bermingham, Alan and others, *The Small Television Studio*, Butterworth, Focal Press (1976)
Bibby, A, *Local Television: Piped Dreams?*, Redwing Press (1979)
Black, P, *The Biggest Aspidistra in the World*, BBC (1972)
Bland, K, *You're on Next! How to Survive on TV and Radio*, Kogan Page (1983)
Bliss, Edward and Patterson, John, *Writing News for Broadcast*, Columbia University Press (1971)
Blue Book of British Broadcasting. A Handbook for Professional Bodies and Students of Broadcasting, Telex Monitors Ltd
Blumler, J G and others, *The Challenge of Election Broadcasting*, Leeds University Press (1978)
Bow Group, *Pay Cable: the TV Revolution that is Coming to GB*, Bow Group (1982)
Boyle, A, *Only the Wind will Listen: Reith of the BBC*, Hutchinson (1972)
Brand, J, *Hello, Good Evening and Welcome - a Guide to being Interviewed on Television and Radio*, Shaw (1984)
Brandt, G W, *British Television Drama*, Cambridge University Press (1981)
Bridson, D G, *Prospero and Ariel. The Rise and Fall of Radio, a Personal Recollection*, Gollancz (1971)
Briggs, A, *The BBC: the First Fifty Years*, Oxford University Press (1985)
Briggs, A, *Governing the BBC*, BBC (1979)
Briggs, S, *Those Radio Times*, Weidenfeld & Nicolson (1981)
Briscoe, D and Curtis-Bramwell, R, *The BBC Radiophonic Workshop, the First 25 Years 1958-83*, BBC (1983)
Brown, Donald and Jones, John, *Radio and Television News*, Hastings House (1972)
Brown, J and Glazier, E V D, *Telecommunications*, Chapman & Hall (1975)
Burns, T, *The BBC: Public Institution and Private World*, Macmillan (1977)

Cable Debate. A BBC Briefing, BBC (1982)
Carrick, Edward, *Designing for Film*, Studio Publications
Cawston, R and others, *Principles and Practice in Documentary Programmes*, BBC (1972)
Ceefax, its History and Development, BBC (1978)
Clutterbuck, R, *The Media and Political Violence*, Macmillan (1983)
Cohen, S and Young, J, *The Manufacture of News: Deviance, Social Problems and the Mass Media*, Constable (1973)
COI, *Broadcasting in Britain*, Central Office of Information (1981)
COI, *Educational Television and Radio in Britain*, Central Office of Information (1981)
Conrad, P, *Television: the Medium and its Manners*, Routledge & Kegan Paul (1982)
Cooke, Brian, *Writing Comedy for Television*, Methuen (1983)
Cox, G, *See it happen: the Making of ITN*, Bodley Head (1983)
Cullingford, C, *Children and Television*, Gower Press (1984)
Curran, C J, ed, *Mass Communication and Society*, Arnold (1977)
Curran, C J and Seaton, J, *Power without Responsibility: the Press and Broadcasting in Britain*, Fontana (1981)
Davis, Desmond, *The Grammar of Television Production*, Barrie & Jenkins (1978)
Day, Robin, *Day by Day*, Kimber (1975)
Development of Cable Systems and Services, HMSO (1983)
Dimbleby, J, *Richard Dimbleby*, Hodder & Stoughton (1975)
Direct Broadcasting by Satellite: Report of a Home Office Study, HMSO (1981)
Drakakis, J, ed, *British Radio Drama*, Cambridge University Press (1981)
Ehrenberg, A S C, *The Funding of BBC Television*, London Business School (1984)
Electronic Media Directory and New Media Yearbook, WOAC Communications Company (annual)
Englander, A A, *Filming for Television*, Butterworth, Focal Press (1976)
Esslin, M, *The Age of Television*, W H Freeman (1982)
Evans, E, *Radio: a Guide to Broadcasting Techniques*, Barrie & Jenkins (1977)
Eysenck, H J, *Sex, Violence and the Media*, Paladin (1982)
Falconer, R, *Message, Media, Mission*, St Andrew Press (1977)
Fane, Irving E, *Television News*, Butterworth, Focal Press (1972)
Field, Syd, *Screenplay ... A step-by-step Guide from Concept to Finished Script*, Delacorte Press (nd)
Fisher, D and Tasker, J, eds, *Education and Training for Film and Television*, British Kinematography, Sound and Television Society (1977)
Foss, H, ed, *Video Production Techniques*, Kluwer (1980)
Francis, R, *Television - the Evil Eye?*, BBC (1981)
Francis, R, *What Price Free Speech?*, BBC (1982)
Gabriel, Jim, *Thinking about Television*, Oxford University Press (nd)
Gagliardi, R M, *Introduction to Communications Engineering*, Wiley (1978)

Gilford, C, *Acoustics for Radio and TV Studios*, P Peregrinus for IEE (1972)
Glasgow University Media Group, *Bad News*, Routledge & Kegan Paul (1976)
Glasgow University Media Group, *More Bad News*, Routledge & Kegan Paul (1980)
Glasgow University Media Group, *Really Bad News*, Writers' and Readers' Publishing Co-operative Society (1982)
Goldie, Grace Wyndham, *Facing the Nation. Television and Politics 1936-76*, Bodley Head (1977)
Golding, P, *The Mass Media*, Longman (1974)
Golding, P and Elliott, P, *Making the News*, Longman (1979)
Goodhardt, G J and others, *The Television Audience: Patterns of Viewing*, Saxon House (1975)
Graham J, ed, *Penguin Dictionary of Telecommunications*, A Lane, Penguin (1983)
Guide to Acoustic Practice, BBC (1980)
Gunter, B, *Dimensions of Television Violence*, Gower Press (1984)
Hall, Mark, *Broadcast Journalism: an Introduction to News Writing*, Hastings House (1971)
Happe, B, *Basic Motion Picture Technology*, Butterworth, Focal Press (1978)
Happe, B, *Dictionary of Audio-Visual Terms*, Butterworth, Focal Press (1983)
Hartley, I, *Goodnight Children ... everywhere*, Midas Books (1983)
Hartley, J, *Understanding News*, Methuen (1982)
Hawkridge, D and Robinson, J, *Organising Educational Broadcasting*, Croom Helm (1981)
Hawthorne, J, *Reporting Violence: Lessons from Northern Ireland?*, BBC (1981)
Haykin, S, *Communication Systems*, Wiley (1984)
Hearst, S, *Artistic Heritage and its Treatment by Television*, BBC (1981)
Herbert, John, *The Techniques of Radio Journalism*, A & C Black (1976)
Herdeg, Walter and Hales, John, *Film and Television Graphics*, Visual Communications Books (nd)
Hill, C, *Behind the Screen: the Broadcasting Memoirs of Lord Hill*, Sidgwick & Jackson (1974)
Hillard, Robert L, *Radio Broadcasting*, Butterworth, Focal Press (1975)
Hillard, Robert L, *Writing for Television and Radio*, Butterworth, Focal Press (1976)
Hoggart, R and Morgan J, eds, *The Future of Broadcasting: Essays on Authority, Style and Choice*, Macmillan (1982)
Hollins, T, *Beyond Broadcasting: into the Cable Age*, BFI Publishing for the Broadcasting Research Unit (1984)
Hood, Stuart, *On Television*, Pluto Press (1983)
Hood, Stuart, *The Professions - Radio and Television*, David & Charles (nd)
Horner, R, *Inside BBC Television: a Year behind the Camera*, Webb & Bower: BBC TV (1983)
Howard, G, *BBC Educational Broadcasting and the Future*, BBC (1981)
Howard, G, *Towards 1996*, BBC (1981)

Howe, M J A, *Television and Children*, New University Education (1977)
Hunt, A, *The Language of Television: Uses and Abuses*, Eyre Methuen (1981)
Hurrell, Ron, *The Thames & Hudson Manual of Television Graphics*, Thames & Hudson (1973)
IBA *Television and Radio*, Independent Broadcasting Authority (yearbook)
IBA *Annual Report and Accounts*, available at HMSO
IBA Code of Advertising Standards and Practice (1982)
IBA/BBC *The Portrayal of Violence on Television, IBA/BBC Guidelines*, IBA/BBC (1983)
Jones, Peter, *The Technique of the Television Cameraman*, Butterworth, Focal Press (nd)
Jones, Peter, *Television behind the Scenes*, Blandford Press (1984)
Kehoe, Vincent, *The Techniques of Film and Television Makeup*, Butterworth, Focal Press
Kemps International Film and Television Yearbook, (annual)
Knightley, Philip, *The First Casualty: the War Correspondent as Hero ...*, Deutsch (1975)
Kuhn, R, ed, *The Politics of Broadcasting*, Croom Helm (1985)
Lambert, S, *Channel Four: TV with a Difference?*, BFI (1982)
Large, M, *Who's Bringing Them Up? Television and Child Development*, TV Action Group (1980)
Laughton, Roy, *TV Graphics*, Vista (1966)
Leeming, Jan, *Working in Television*, Batsford (nd)
Lewis, John and Brinkley, *Graphic Design*, Routledge & Kegan Paul (nd)
Lewis, P F, *Radio Drama*, Longman (1981)
Lewis, P M, *Whose Media? The Annan Report and After*, Consumers' Association (1978)
Lyle, Garry, *Broadcasting*, (1973)
McLeish, R, *The Technique of Radio Production*, Butterworth, Focal Press (1978)
McNae, L J C and Taylor, R M, *Essential Law for Journalists*, Staples Press (1972)
Magee, Bryan, *The Television Interviewer*, Macdonald (1966)
Mansell, G, *Broadcasting to the World: Forty Years of BBC External Services*, BBC (1973)
Mansell, G, *Let Truth be Told: Fifty Years of BBC External Broadcasting*, Weidenfeld & Nicolson (1982)
Millerson, Gerald, *Basic TV Staging*, Butterworth, Focal Press (1982)
Millerson, Gerald, *The Technique of Lighting for Television and Motion Pictures*, Butterworth, Focal Press (1982)
Millerson, Gerald, *TV Lighting Methods*, Butterworth, Focal Press (1982)
Millerson, Gerald, *Effective TV Production*, Butterworth, Focal Press (1983)
Millerson, Gerald, *The Technique of Television Production*, Butterworth, Focal Press (1985)
Milton, Ralph, *Radio Programming, a Basic Training Manual*, Collins (1968)
Mitchell, Leslie, *Leslie Mitchell Reporting*, Hutchinson (1981)
Moorfoot, R, *Television in the Eighties: the Total Equation*, BBC (1982)

Morley, D and Whitaker, B, eds, *The Press, Radio and Television: an Introduction to the Media*, Comedia (1983)
Muggeridge, D, *The New War of the Airwaves*, BBC (1983)
Newby, P H, *The Uses of Broadcasting*, BBC (1978)
Nisbett, A, *The Sound Studio*, Butterworth, Focal Press
Nisbett, A, *The Technique of the Sound Studio: for Radio, Television and Film*, Butterworth, Focal Press (1979)
Paice, E, *The Way to Write for Television*, Elm Tree Books (1981)
Parker, B and Farrell, N, *Television and Radio: Everybody's Soapbox*, Blandford Press (1983)
Parker, D, *Radio: the Great Years*, David & Charles (1977)
Paulu, B, *Television and Radio in the United Kingdom*, Macmillan (1981)
Pegg, M, *Broadcasting and Society 1918–39*, Croom Helm (1983)
Piepe, A, *Mass Media and Cultural Relationships*, Saxon House (1978)
Pike, Frank, ed, *Ah! Mischief – the Writer and Television* ..., Faber & Faber (1982)
Pitt, Hugh, *Television*, Hale (nd)
Prothero, A, *Holding the Balance in Current Affairs Programmes* ..., BBC (1983)
Public for the Visual Arts in Britain, BBC (1980)
Redfern, B, *Local Radio*, Butterworth, Focal Press (1979)
Reith, J C W, *Into the Wind*, Hodder & Stoughton (1949)
Reith, J C W, *Diaries* (ed C Stuart), Collins (1975)
Report of the Radio Network Working Party, BBC (1981)
Robinson, J, *Learning over the Air: Sixty Years of Partnership in Adult Learning*, BBC (1982)
Robinson, J F and Beards, P H, *Using Videotape*, Butterworth, Focal Press (1981)
Robinson, J F, *Videotape Recording: Theory and Practice*, Butterworth, Focal Press (1981)
Roberts, R S, *Dictionary of Audio, Radio and Video*, Butterworth (1981)
Rodger, I, *Radio Drama*, Macmillan (1982)
Rogers, J, *Using Broadcasting with Adults*, BBC (1980)
Rowlands, A, *Script Continuity and the Production Secretary in Film and TV*, Butterworth, Focal Press (1983)
Sandbank, C P, *Optical Fibre Communication Systems*, Wiley (1980)
Schlesinger, P, *Putting 'Reality' Together: BBC News*, Constable (1978)
Science Museum, *Telecommunications: a Technology for Change*, HMSO (1983)
Scroggie, M G, *Foundations of Wireless and Electronics*, Newnes, 10th edition (1984)
See, David, *How to be a DJ*, Hamlyn (1980)
Seglow, P, *Trade Unionism in Television*, Saxon House (1978)
Sendall, B, *Independent Television in Britain*, 2 vols, Macmillan (1982, 3)
Silvey, Robert, *Who's Listening?*, Hamlyn (1980)
Sims, H V, *Principles of PAL Colour Television and Related Systems*, Newnes (1976)
Singer, Aubrey, *Science Broadcasting*, BBC (1966)

Smith, A, *British Broadcasting*, David & Charles (1974)
Smith, A, *The Politics of Information: Problems of Policy in Modern Media*, Macmillan (1978)
Snagge, J and Barnsley, Michael, *Those Vintage Years of Radio*, Pitman (1972)
Sproson, W N, *Colour Science in TV and Display Systems*, Adam Hilger (1982)
Sutton, S, *The Largest Theatre in the World: 30 Years of TV Drama*, BBC (1982)
Swain, D V, *Scripting for Video and Audio Visual Media*, Butterworth, Focal Press (1981)
Syfret, T, *Cable and Advertising in the Eighties*, J Walter Thompson (1983)
Taylor, Glenhill, *Before Television: the Radio Years*, Barnes Yoseloff (1979)
Timpson, John, *Today and Yesterday*, Allen & Unwin (1976)
Timpson, John, *The Lighter Side of Today*, Allen & Unwin (1983)
Took, B, *Laughter in the Air: an Informal History of British Radio Comedy*, Robinson and BBC (1982)
Townsend, Boris, *PAL Colour Television*, Institute of Electrical Engineers (nd)
Tracy, Michael, *The Production of Political Television*, Routledge & Kegan Paul (1978)
Tracy, Michael, *A Variety of Lives: a Biography of Sir Hugh Greene*, Bodley Head (1983)
Trethowan, I, *Broadcasting and Politics*, BBC (1977)
Trethowan, I, *Broadcasting and Society*, BBC (1981)
Trethowan, I, *The Next Age of Broadcasting*, Conservative Political Centre (1984)
Trethowan, I, *Split Screen*, Hamish Hamilton (1984)
Tumber, H, *Television and the Riots*, British Film Institute (1982)
Tunstall, Jeremy, *The Media in Britain*, Constable (1983)
25 Years on ITV 1955-80, ITV Books and Michael Joseph (1980)
Tyrrell, R, *The Work of the Television Journalist*, Butterworth, Focal Press (1980)
Veljanovski, C G and Bishop, W D, *Choice by Cable: the Economics of a New Era in Television*, Institute of Economic Affairs (1983)
Watts, Harris, *On Camera: How to Produce Film and Video*, BBC (1984)
Wedell, E G, *Structures of Broadcasting: a Symposium*, Manchester University Press (1970)
Wenham, B, ed, *The Third Age of Broadcasting*, Faber (1982)
Whale, J, *The Politics of the Media*, Manchester University Press (1977)
White, G, *Video Techniques*, Newnes (1982)
Wilkie, Bernard, *Technique of Special Effects in Television*, Butterworth, Focal Press (1971)
Windlesham, Lord, *Broadcasting in a Free Society*, Blackwell (1980)
Writers' and Artists' Yearbook, A & C Black (annual)
Writing for the BBC..., BBC (1983)
Yorke, Ivor, *The Technique of Television News*, Butterworth, Focal Press (1978)
Young, B, *The IBA and Channel Four*, IBA (1979)

Young, B, *The Paternal Tradition in British Broadcasting 1922-?*, The Watt Club Lecture, Heriot-Watt University (1983)

Periodicals

AIP & Co (the magazine for members of the Association of Independent Producers)
Airwaves
Ariel
Audio Visual Directory
Broadcast
Broadcast Sound
Broadcast Yearbook and Diary
Cable and Broadcast
Cablevision News
Campaign
The Listener
Look-in
Marketing Directory
Media, Culture and Society
Media International
The Media Reporter
Radio Communication
Radio Times
Sight and Sound
Stage and Television Today
TV Times
UK Press Gazette

Chapter 6
Useful Addresses

BBC

Corporate Headquarters and Radio
Broadcasting House, BBC London W1A 1AA; 01-580 4468

Television
Television Centre, Wood Lane, London W12 7RJ; 01-743 8000

External Broadcasting
PO Box 76, Bush House, Strand, London WC2B 4PH; 01-240 3456

BBC/Open University Production Centre
Walton Hall, Milton Keynes MK7 6BH; 0908 655335

Northern Ireland
Broadcasting House, 25–27 Ormeau Avenue, Belfast BT2 8HQ; 0232 244400

BBC Radio Foyle
PO Box 927, Londonderry; 0504 262244/5/6

Scotland
Broadcasting House
Queen Margaret Drive, Glasgow G12 8DG; 041-330 2345

Broadcasting House
5 Queen Street, Edinburgh EH2 1JF; 031-225 3131

Broadcasting House
Beechgrove Terrace, Aberdeen AB9 2ZT; 0224 635233

Broadcasting House
12–13 Dock Street, Dundee; 0382 25025/25905

BBC Radio Highland
7 Culdethel Road, Inverness IV2 4AD; 0463 221711

BBC Radio nan Eilean
Rosebank, Church Street, Stornoway; 0851 5000

BBC Radio Orkney
Castle Street, Kirkwall; 0856 3939

BBC Radio Shetland
Brentham House, Lerwick, Shetland ZE1 0LR; 0595 4747

BBC Radio Solway
Elmbank, Lovers' Walk, Dumfries DG1 1NZ; 0387 68008

BBC Radio Tweed
Municipal Buildings, High Street, Selkirk TD7 4BU; 0750 21884

Wales
Broadcasting House
Llantrisant Road, Llandaff, Cardiff CF5 2YQ; 0222 564888

Broadcasting House
Meirion Road, Bangor LL57 2BY; 0248 362214

Broadcasting House
32 Alexandra Road, Swansea SA1 5DZ; 0792 54986

BBC Radio Clwyd
The Old School House, Glanrafon Road, Mold CH7 1PA; 0352 59111

BBC Radio Gwent
Powys House, Cwmbran, Gwent NP44 1YF; 06333 72727

BBC Local Radio Stations
BBC Radio Bedfordshire
PO Box 476, Hastings Street, Luton, Bedfordshire LU1 5BA; 0582 459111

BBC Radio Bristol
3 Tyndalls Park Road, Bristol BS8 1PP; 0272 74111

BBC Radio Cambridgeshire
Broadcasting House, Hills Road, Cambridge CB2 1LD; 0223 315970

BBC Radio Cleveland
PO Box 1548, Broadcasting House, Newport Road, Middlesbrough, Cleveland TS1 5DG; 0642 225211

BBC Radio Cornwall
Phoenix Wharf, Truro, Cornwall TR1 1UA;
0872 75421

BBC Radio Cumbria
Hilltop Heights, London Road, Carlisle, Cumbria CA1 2NA; 0228 31661

BBC Radio Derby
56 St Helen's Street, Derby DE1 3HY; 0332 361111

BBC Radio Devon
PO Box 100, St David's Hill, Exeter, Devon EX4 4DB; 0392 215651

BBC Radio Furness
(Radio Cumbria community opt-out station)
Broadcasting House, Hartington Street, Barrow-in-Furness, Cumbria LA14 5SH; 0229 36767

BBC Radio Guernsey
Commerce House, Les Banques, St Peter Port, Guernsey, Channel Islands; 0481 28977

BBC Radio Humberside
63 Jameson Street, Hull HU1 3NU; 0482 23232

BBC Radio Jersey
Broadcasting House, Rouge Bouillon, St Helier, Jersey, Channel Islands; 0534 70000

BBC Radio Kent
30 High Street, Chatham, Kent ME4 4EZ; 0634 46284

BBC Radio Lancashire
King Street, Blackburn, Lancashire BB2 2EA; 0254 62411

BBC Radio Leeds
Broadcasting House, Woodhouse Lane, Leeds LS2 9PN; 0532 442131

BBC Radio Leicester
Epic House, Charles Street, Leicester LE1 3SH; 0533 27113

BBC Radio Lincolnshire
Radion Buildings, PO Box 219, Newport, Lincoln LN1 3DF; 0522 40011

BBC Radio London
35a Marylebone High Street, London W1A 4LG; 01-486 7611

BBC Radio Manchester
New Broadcasting House, PO Box 90, Oxford Road, Manchester M60 1SJ; 061-228 3434

BBC Radio Merseyside
55 Paradise Street, Liverpool L1 3BP; 051-708 5500

BBC Radio Newcastle
Crestina House, Archbold Terrace, Newcastle upon Tyne NE2 1DZ; 091-281 4243

BBC Radio Norfolk
Norfolk Tower, Surrey Street, Norwich NR1 3PA; 0603 617411

BBC Radio Northampton
PO Box 1107 Abington Street, Northampton NN1 2BE; 0604 20621

BBC Radio Nottingham
York House, Mansfield Road, Nottingham NG1 3JB; 0602 415161

BBC Radio Oxford
242–254 Banbury Road, Oxford OX2 7DW; 0865 53411

BBC Radio Sheffield
Ashdell Grove, 60 Westbourne Road, Sheffield S10 2QU; 0742 686185

BBC Radio Shropshire
2–4 Boscobel Drive, Shrewsbury, Shropshire SY1 3TT; 0743 248484

BBC Radio Solent

South Western House, Canute Road, Southampton SO9 4PJ; 0703 31311

BBC Radio Stoke-on-Trent
Conway House, Cheapside, Hanley, Stoke-on-Trent, Staffordshire ST1 1JJ; 0782 24827

BBC Radio Sussex
Marlborough Place, Brighton, Sussex BN1 1TU; 0273 680231

BBC WM (West Midlands)
PO Box 206, Pebble Mill Road, Birmingham B5 7SD; 021-472 5141

BBC Radio York
20 Bootham Row, York YO3 7BR; 0904 641351

Network Production Centres
Pebble Mill, Pebble Mill Road, Birmingham B5 7QQ; 021-472 5353

Broadcasting House, Whiteladies Road, Clifton, Bristol BS8 2LR; 0272 732211

New Broadcasting House, Oxford Road, Manchester M60 1SJ; 061-236 8444

Regional Television Stations
English Regional Headquarters
Pebble Mill, Pebble Mill Road, Birmingham B5 7QQ; 021-472 5353

BBC East
St Catherine's Close, All Saints Green, Norwich NR1 3ND; 0603 28841

BBC Midlands
Broadcasting Centre, Pebble Mill Road, Birmingham B5 7QQ; 021-472 5353

BBC North
Broadcasting Centre, Woodhouse Lane, Leeds LS2 9PX; 0532 41181

BBC North-East
Broadcasting House, 54 New Bridge Street, Newcastle upon Tyne NE1 8AA; 0632 320961

BBC North-West
New Broadcasting House, Oxford Road, Manchester M60 1SJ; 061-236 8444

BBC South
Southwestern House, Canute Road, Southampton SO9 1PF; 0703 26201

BBC South-West
Broadcasting House, Seymour Road, Mannamead, Plymouth PL3 5DB; 0752 29201

BBC West
Broadcasting House, 21-33b Whiteladies Road, Clifton, Bristol BS8 2LR; 0272 732211

Independent Local Radio (in alphabetical order of area)

Northsound Radio
45 Kings Gate, Aberdeen AB2 6BL; 0224 632234

West Sound
Radio House, 54 Holmston Road, Ayr KA7 3BE; 0292 283662

Chiltern Radio (Luton/Bedford)
Chiltern Road, Dunstable, Bedfordshire LU6 1HQ; 0582 666001
and at
55 Goldington Road, Bedford MK40 3LS; 0234 49266

Downtown Radio (Belfast)
PO Box 96, Newtownards BT23 4ES, Northern Ireland; 0247 815555

BRMB Radio
Radio House, PO Box 555, Aston Road North, Birmingham B6 4BX; 021-359 4481

2CR (Two Counties Radio)
5-7 Southcote Road, Bournemouth BH1 3LR; 0202 294881

Pennine Radio
PO Box 235, Pennine House, Forster Square, Bradford BD1 5NP; 0274 731521

Southern Sound
Radio House, Franklin Road, Portslade, Brighton BN4 2SS; 0273 422288

GWR
PO Box 2000, Watershed, Canon's Road, Bristol BS99 7SN; 0272 279900

Saxon Radio
Long Brackland, Bury St Edmunds, Suffolk IP33 1JY; 0284 701511

Invicta Radio (Maidstone and Medway)
(incorporating Northdown Radio and Network East Kent)
15 Station Road East, Canterbury CT1 2RB; 0227 67661

Red Dragon Radio
Radio House, West Central Wharf, Cardiff CF1 5XI; 0222 384041

Mercia Sound
Hertford Place, Coventry CV1 3TT; 0203 28451

Radio Mercury
Broadfield House, Brighton Road, Crawley, West Sussex RH11 9TT; 0293 519161

Radio Tay
PO Box 123, Dundee DD1 9UF; 0382 29551

Radio Forth
Forth House, Forth Street, Edinburgh EH1 3LF; 031-556 9255

Devon Air Radio
35-37 St David's Hill, Exeter EX4 4DA; 0392 30703

Radio Clyde
Clydebank Business Park, Clydebank, Glasgow G81 2RX; 041-941 1111

Severn Sound
Old Talbot House, 67 Southgate Street, Gloucester GL1 2DO;
0452 423791

County Sound
The Friary, Guildford GU1 4YX; 0483 505566

Radio Wyvern (Hereford/Worcester)
(see Worcester)

Viking Radio
Commercial Road, Hull HU1 2SG; 0482 25141

Moray Firth Radio
PO Box 271, Inverness IV3 6SF; 0463 224433

Radio Orwell
Electric House, Lloyds Avenue, Ipswich IP1 3HZ; 0473 216971

Radio Aire
PO Box 362, Leeds LS3 1LR; 0532 452299

Leicester Sound
Granville House, Granville Road, Leicester LE1 7RW; 0533 551616

Radio City
PO Box 194, Liverpool L69 1LD; 051-227 5100

Capital Radio (general and entertainment)
Euston Tower, London NW1 3DR; 01-388 1288

London Broadcasting Company (LBC)
(news and information)
Communications House, Gough Square, London EC4P 4LP; 01-353 1010

Chiltern Radio (Luton/Bedford)
(See Bedfordshire)

Invicta Radio
(incorporating Northdown Radio and Network East Kent)
37 Earl Street, Maidstone ME14 1PF; 0622 679061

Piccadilly Radio
127–131 The Piazza, Piccadilly Plaza, Manchester M1 4AW; 061-236 9913

Gwent Broadcasting (Newport)
(service currently provided by Red Dragon Radio)

Hereward Radio
PO Box 1557, Abington Street, Northampton NN1 2HW; 0604 29811

Radio Broadland (Great Yarmouth and Norwich)
47–49 St George's Plain, Colegate, Norwich NR3 1DD; 0603 630621

Radio Trent
29-31 Castle Gate, Nottingham NG1 7AP; 0602 581731

Hereward Radio
PO Box 225, 114 Bridge Street, Peterborough PE1 1XJ; 0733 46225

Plymouth Sound
Earl's Acre, Alma Road, Plymouth PL3 4HX; 0752 27272

Radio Victory
PO Box 257, Portsmouth PO1 5RT; 0705 827799

Red Rose Radio
PO Box 301, St Paul's Square, Preston PR1 1YE; 0772 556301

Radio 210 (Thames Valley Broadcasting)
PO Box 210, Reading, Berkshire RG3 5RZ; 0734 413131

Radio Mercury (Reigate and Crawley)
Broadfield House, Brighton Road, Crawley, West Sussex RH11 9TT; 0293 419161

Radio Hallam
PO Box 194, Hartshead, Sheffield S1 1GP; 0742 71188

Essex Radio
Radio House, Clifftown Road, Southend-on-Sea SS1 1SX; 0702 333711

Radio Tees
74 Dovecot Street, Stockton-on-Tees, Cleveland TS18 1HB; 0642 615111

Signal Radio
Studio 257, Stoke Road, Stoke-on-Trent ST4 2SR; 0782 417111

Swansea Sound
Victoria Road, Gowerton, Swansea SA4 3AB; 0792 893751

GWR
PO Box 2000, Lime Kiln Studios, Wootton Bassett, Swindon SN4 7EX; 0793 853222

Metro Radio (Tyne and Wear)
Newcastle upon Tyne NE99 1BB; 091-488 3131

Beacon Radio
PO Box 303, 267 Tettenhall Road, Wolverhampton WV6 0DO; 0902 757211

Radio Wyvern (Hereford and Worcester)
5-6 Barbourne Terrace, Worcester WR1 3JS; 0905 612212

Marcher Sound
Sain-Y-Gororau, The Studios, Gwersylit, Wrexham, Clwyd LL11 4AF; 0978 752202/0244 372202

Independent Television Companies

Anglia Television
Anglia House, Norwich NR1 3JG; 0603 615151

Border Television

Television Centre, Carlisle CA1 3NT; 0228 25101

Central Independent Television
Central House, Broad Street, Birmingham B1 2IP; 021-643 9898

Channel Four Television Company Limited
60 Charlotte Street, London W1P 2AX; 01-631 4444

Channel Four Wales
(see Sianel Pedwar Cymru)

Channel Television
The Television Centre, St Helier, Jersey, Channel Islands; 0534 73999
and at
The Television Centre, St George's Place, St Peter Port, Guernsey, Channel Islands; 0481 23451

Grampian Television
Queen's Cross, Aberdeen AB9 2XI; 0224 646464

Granada Television
Granada TV Centre, Manchester M60 9EA; 061-832 7211

HTV Wales
Television Centre, Colverhouse Cross, Cardiff CF5 6XJ; 0222 590590

HTV West
Television Centre, Bath Road, Bristol BS4 3HG; 0272 778366

Independent Television News (ITN)
ITN House, 48 Wells Street, London W1P 4DE; 01-637 2424

London Weekend Television
South Bank Television Centre, Kent House, Upper Ground, London SE1 9LT; 01-261 3434

Scottish Television
Cowcaddens, Glasgow G2 3PR; 041-332 9999

Sianel Pedwar Cymru
Clos Sophia, Caerdydd CF1 9XY; 0222 43421

Television South
Television Centre, Southampton SO9 5HZ; 0703 3411

Thames Television
Thames Television House, 306–316 Euston Road, London NW1 3BB; 01-387 9494

TSW (Television South West)
Derry's Cross, Plymouth, Devon PL1 2SP; 0752 663322

TV-am
Breakfast Television Centre, Hawley Crescent, London NW1 8EF; 01-267 4300/4377

Tyne Tees Television
The Television Centre, City Road, Newcastle upon Tyne NE1 2AL; 0632 610181

Ulster Television
Havelock House, Ormeau Road, Belfast BT7 1EB; 0232 228122

Yorkshire Television
The Television Centre, Leeds LS3 1JS; 0532 438283

Other Addresses

Advertising, Film and Videotape Producers' Association
48 Carnaby Street, London W1; 01-434 2651

Association of Cinematograph, Television and Allied Technicians
2 Soho Square, London W1; 01-437 8506

Association of Independent Producers
17 Great Pulteney Street, London W1; 01-437 3549

Association of Independent Radio Contractors Ltd
1st Floor, Regina House, 259-269 Old Marylebone Road, London
NW1 5RA; 01-262 6681

British Actors' Equity Association
8 Harley Street, London WC1N 2AB; 01-636 6367

British Film and Television Producers' Association Ltd
Paramount House, 162 Wardour Street, London W1; 01-437 7700

British Forces Broadcasting Service
(see Services Sound and Vision)

British Kinematography, Sound and Television Society
110-112 Victoria House, Vernon Place, London WC1B 4DJ; 01-242 8400

British Satellite Broadcasting
19 Rathbone Place, London W1P 1DF; 01-583 1544

Cable Television Association
295 Regent Street, London W1; 01-637 4591

Cable Television Authority
Gillingham House, 38-44 Gillingham Street, London SW1V 1HV;
01-821 6161

Film and Video Press Writers and Editors
37 Gower Street, London WC1E 6HH; 01-580 2842

The Home Office
Queen Anne's Gate, London SW1H 9AT; 01-213 3000

Independent Broadcasting Authority
70 Brompton Road, London SW3 1EY; 01-584 7011

Independent Programme Producers' Association
7 Fitzroy Square, London W1; 01-388 1234

Independent Television Companies Association Ltd
Knighton House, 56 Mortimer Street, Londonn W1N 8AN; 01-636 6866

Independent Television Publications
247 Tottenham Court Road, London W1P 0AU; 01-323 3222

JOBFIT
The Administrator, JOBFIT
Fourth Floor, 5 Dean Street, London W1V 5RN; 01-734 5141

Joint Advisory Committee for Radio Journalism Training
46 Southway, Hampstead, London NW3; 01-485 3199

Manx Radio
PO Box 219, Broadcasting House, Douglas, Isle of Man; 0624 73277

National Association of Hospital Broadcasting Organisations
(Secretary: P Milward), 107 Bare Lane, Morecambe, Lancashire;
0524 415809

National Council for the Training of Journalists
Carlton House, Hemnall Street, Epping, Essex CM16 4NL; 0378 72395

The Newspaper Society (training department)
Whitefriars House, 6 Carmelite Street, London EC4 0BL; 01-583 3311

ORACLE Teletext Ltd
Craven House, 25-32 Marshall Street, London W1V 1LL; 01-434 3121

Radio Marketing Bureau
Regina House, 259-269 Old Marylebone Road, London NW1 5RA;
01-258 3705

Radio Luxembourg (London) Ltd
38 Hertford Street, London W1Y 8BA; 01-493 5961

Radio Writers' Association
84 Drayton Gardens, London SW10 9FB; 01-373 6642

Royal Television Society
Tavistock House East, Tavistock Square, London WC1H 9HR;
01-387 1970

Services Sound and Vision Corporation
Bridge House, North Wharf Road, London W2 1LA; 01-724 1234

Television Production Companies

For a list of names and addresses consult *Kemps International Film and Television Yearbook* and the *Electronic Media Directory and New Media Yearbook*, WOAC Communications Company, 10 Friars Walk, Dunstable, Bedfordshire LU6 3SA; 0582 660138.

The Kogan Page Careers Series

This series consists of short guides (96-160 pages) to different careers for school-leavers, graduates and anyone wanting to start anew. Each book serves as an introduction to a particular career and to jobs available within that field, including full details of training qualifications and courses. The following 'Careers in' titles are available in paperback.

Accountancy *(2nd edition)*
Alternative Medicine
Antiques
Art and Design *(4th edition)*
Aviation
Banking and Finance
Business
Catering and Hotel Management *(2nd edition)*
The Church
Civil Engineering
Civil Service
Computing and Information Technology
Conservation *(2nd edition)*
Crafts
Dance
Electrical and Electronic Engineering *(2nd edition)*
Engineering *(3rd edition)*
Fashion
Floristry and Retail Gardening
Hairdressing and Beauty Therapy *(3rd edition)*
Holiday Industry
Home Economics *(2nd edition)*
Journalism *(2nd edition)*
Land and Property
The Law *(2nd edition)*
Librarianship and Information Science *(2nd edition)*
Marketing, Public Relations and Advertising *(2nd edition)*
Medicine, Dentistry and Mental Health *(3rd edition)*
Modelling *(2nd edition)*
Museums and Art Galleries
Music Business
Nursing and Allied Professions *(3rd edition)*
Oil and Gas
Pharmacy
Photography *(2nd edition)*
The Police Force *(2nd edition)*
Politics
Printing
Psychology
Publishing
Retailing *(2nd edition)*
Road Transport
At Sea
Secretarial and Office Work *(2nd edition)*
Social Work *(3rd edition)*
Sport *(2nd edition)*
Surveying
Teaching *(2nd edition)*
Telecommunications
Television and Radio *(2nd edition)*
The Theatre *(2nd edition)*
Using Biology
Using Languages *(2nd edition)*
Using Mathematics
Veterinary Surgery
Working Abroad
Working Outdoors *(3rd edition)*
Working with Animals *(3rd edition)*
Working with Children and Young People *(3rd edition)*